Heba Abo Ebtehan

Functionalized Amine Mesoporous Silica SBA-15 Supporting Metal Oxides

Heba Abo Ebtehan

Functionalized Amine Mesoporous Silica SBA-15 Supporting Metal Oxides

Noor Publishing

Imprint

Any brand names and product names mentioned in this book are subject to trademark, brand or patent protection and are trademarks or registered trademarks of their respective holders. The use of brand names, product names, common names, trade names, product descriptions etc. even without a particular marking in this work is in no way to be construed to mean that such names may be regarded as unrestricted in respect of trademark and brand protection legislation and could thus be used by anyone.

Cover image: Provided by the author

Publisher:
Noor Publishing
is a trademark of
International Book Market Service Ltd., member of OmniScriptum Publishing Group
17 Meldrum Street, Beau Bassin 71504, Mauritius

Printed at: see last page
ISBN: 978-620-0-06691-6

Functionalized Amine Mesoporous Silica SBA-15 Supporting Metal Oxides Nanoparticles: Synthesis and Characterization

Heba Abed Allah Abu Ebtihan

﴿قَالُواْ سُبْحَانَكَ لَا عِلْمَ لَنَا إِلَّا مَا عَلَّمْتَنَا إِنَّكَ أَنتَ الْعَلِيمُ الْحَكِيمُ﴾

صدق الله العظيم

Dedication

To my parents,

Dear children
(Saleh, Mena, Dema, Sallah),

Brothers and sisters .

Acknowledgements

First of all, I am grateful to Allah, who granted me the power and courage to finish this study.

I would like to thank my supervisors, Prof. **Issa M. El-Nahhal** *,Prof.* **Jamil K. Salem** *for their great support, enthusiasm, and guidance over the last year. I had the privilege of being involved in many interesting scientific discussions, which tremendously influenced and improved my scientific work. Also I would like to thank prof.* **Sh. Zourob** *and prof.* **T. Hammad** *for their discussions.*

My deepest appreciation and sincere gratitude, to my husband and sisters, who helped me at various stages during my work.

I also extend my profound thanks, to the department of chemistry in Al-Azhar University of Gaza, and the staff of chemistry department in Al-Aqsa University of Gaza ,for making the facilities available for me to use.

Heba A. Abu Ebtihan.

Abstract

Synthesis of mesoporous silica using surfactant templating leads to highly ordered pore distribution, its supporting by metal oxide nanoparticles give highly catalytic activity. In this work mesoporous SBA-15 silica was synthesized using triblockcopolymers P123. Zinc Oxide (ZnO) and Copper Oxide (CuO) nanoparticles with different percentages (5,10,15 and 20%),were successfully loaded into mesoporous silica SBA-15, by impregnation method. Amine functional group was grafted onto mesoporous silica SBA-15 modified with metal oxides nanoparticles, by post condensation method. Several techniques were used for structural examination of these materials like, fourier transform infra red spectroscopy (FT-IR)spectroscopy, wide and small angle X-ray diffraction (XRD), thermal gravimetric analysis (TGA) and photoluminescence spectroscopy (PL). FT-IR spectra showed that, the loaded metal oxide were not chemically interacted with the host mesoporous silica SBA-15 and probably physically interacted with silica network. XRD analysis showed that, the loaded metal oxides were found in the crystalline form as hexagonal structure for ZnO and as monoclinic structure for CuO. The properties of mesoporous silica SBA-15 do not changed by the introduction of metal oxide nanoparticles, and only slight decreasing in the d-spacing after loading of metal oxide, was confirmed by XRD small angle. TGA analysis had approved that mono amine functional groups were grafted onto surface of mesoporous silica coated metal oxide nanoparticles. The amine functionalized mesoporous silica SBA-15, immobilized and free immobilized ZnO (20%) , showed a removal of E124 azo dye from water. UV-vis spectra have indicated that ,these materials exhibited high potential for extraction of toxic dyes.

تحضير ودراسة خواص الميزوبورس سيليكا (SBA-15) المرتبطة وظيفيا بمجموعة الأمين، والمطعمة بجسيمات أكاسيد المعادن.

الملخص العربي

لقد تم تحضير مادة الميزوبورس سيليكا SBA-15 ومن ثم تم إدخال كل من أكسيد الزنك ZnO ، وأكسيد النحاسCuO بنسب مئوية مختلفة (% 20 ,15 ,5,10) داخل فجوات مادة الميزوبورس سيليكا (SBA-15) بطريقة التطعيم Impregnation method . كما تم تركيب مجموعة الأمين الوظيفية على سطح الميزوبورس سيليكا والميزوبورس سيليكا المدعمة بجسيمات أكاسيد المعادن (ZnO, CuO)، بنسبة % 20 باستخدام (Amino silane coupling agent) تبعا لطريقة بوست (Post grafting method).

وقد استخدمت الطرق التحليلية المختلفة لدراسة خواص الميزوبورس سيليكا، والتغيرات التي طرأت عليها بعد إدخال جسيمات أكاسيد المعادن داخل فجواتها، وكذلك عند ارتباطها بمجموعات الأمين الأحادية.

ومنها الأشعة تحت الحمراء FT-IR، التحاليل الحرارية TGA وحيود الأشعة السينية X-RD بزاوية صغيرة وزاوية كبيرة.

وقد أوضحت تحاليل الأشعة تحت الحمراء (FT–IR) أن جسيمات أكاسيد المعادن (ZnO, CuO) تم إدخالها في فجوات مادة SBA–15. وارتباطها كان فيزيائيا وليس كيميائيا.

و من ناحية أخرى فإن حيود الأشعة السينية بزاوية كبيرة أظهرت أن أكسيد الزنك ZnO ظهر في الشكل البلوري السداسي (hexagonal) بينما أكسيد النحاس CuO ظهر على شكل البلوري أحادي الميل (monoclinic).

أما حيود الأشعة السينية بزاوية صغيرة فقد أشارت إلى أن خواص الميزوبورس سيليكا (SBA–15) وشكلها البلوري لم يتغير عند إضافة أكاسيد المعادن، غير انه ظهرت إزاحة طفيفة في المسافة بين الطبقات (d-Spacing).

بينما تحاليل TGA أكدت أن عملية ارتباط مجموعات الأمين الأحادية على سطح الميزوبورس سيليكا والميزوبورس سيليكا المدعمة بأكاسيد المعادن قد تمت بنجاح.

إن هذه المواد لها قدرة على التخلص من الصبغات السامة فقد استخدمت الميزوبورس سيليكا SBA-15 المرتبطة وظيفيا بمجموعة الأمين و المدعمة بأكسيد الزنك والحرة منه بنسبة (20%) للتخلص من صبغة الطعام الحمراء E124 من المحاليل المائية. وأظهرت النتائج أن المواد الثلاث لها القدرة على امتزاز الصبغة وأن الميزوسيليكاSBA-15 الحرة من أكسيد الزنك والمرتبطة بمجموعة الأمين الأحادية لها قدرة أعلى وأسرع في امتزاز الصبغة.

CONTENTS

List of Tables

List of Schemes

List of Figures

List of Abbreviations

Num.	Abbreviation	Full Name
1.	SBA-15	Santa Barbara Amorphous
2.	MCM	Mobil Crystalline Materials
3.	LCT	Liquid Crystal Templating
4.	CTAB	Cetyl Trimethyl Ammonium Bromide
5.	CMC	Critical Micelle Concentration
6.	CVD	Chemical Vapor Deposition
7.	PLD	Pulsed Laser Deposition
8.	NPs	NanoParticles
9.	CPE	Carbon Paste Electrode
10.	RT	Room Temperature
11.	FT-IR	Fourier Transform Infra-Red
12.	TGA	Thermal Gravimetric Analysis
13.	XRD	X-Ray Diffraction
14.	SAXS	Small Angle X-Ray Scattering
15.	WAXS	Wide Angle X-Ray Scattering
16.	SEM	Scanning Electron Microscopy
17.	TEM	Transmission Electron Microscopy
18.	PL	Photoluminescence Spectroscopy
19.	DTA	Differential Thermo gravimetric Analysis
20.	ADI	Acceptable Daily Intake
21.	WHO	World Health Organization
22.	IUBAC	International Union of Pure and Applied Chemistry
23.	BET	Brunauer Emmett Teller

CHAPTER ONE

Introduction

Introduction

Porous materials is very important materials witch use in various fields, such as adsorption, separation, catalysis and sensor. Zeolites (crystalline aluminosilicates with periodic three-dimensional framework structures) with pore size less than 1.2 nm, un-preferred in catalytic processes, the reagents and products could not easily pass through the pores. So the researcher have searched for a new porous materials. In the early 1970s,French scientists develop a method using silica gels and long chain cationic surfactants[1], but the patent didn't produce enough attention, menially due to the lack of XRD and electron microscopy characterization data. Japanese scientists earlier than 1990,also started the synthesis of mesoporous materials, they used cationic surfactant, but it was destroyed in a high alkalinity solutions[1] . Actually in 1992s Mobil Oil Corporation (Mobil) scientists, have reported the first mesoporous silica materials, with ordered pore arrays M41S series ,they not only developed family of mesoporous materials with ordered pore arrangements, but also proposed a general liquid-crystal mechanism [1], with detailed synthesis method. Anew inorganic synthetic chemistry research area began to rise. The porous materials are classified by IUPAC into three categories, depending on diameter of the pore. Microporous materials having pore diameter less than 2nm, macroporous materials, having pore sizes exceeding 50.0 nm, mesoporouse materials having pore sizes intermediate between (2.0-50nm) as shown in Fig.1.1 .

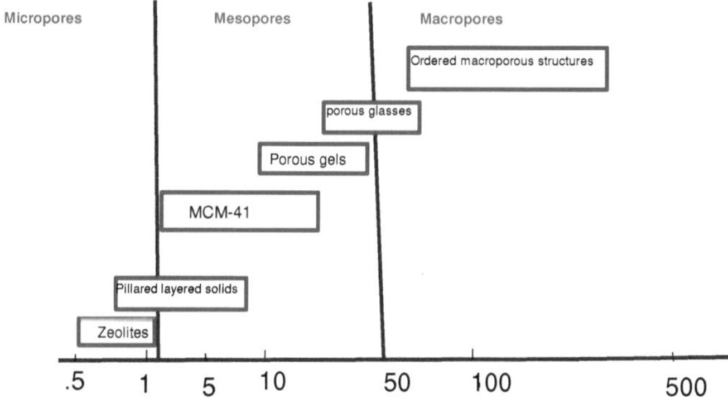

Fig. 1.1 Pore size distributions of some porous materials

1.1 Ordered mesoporous materials

Meso, the Greek prefix, meaning—in between, has been adopted by IUPAC to define porous materials with pore sizes between 2.0 and 50.0 nm [2]. Crystalline mesoporouse materials are very important because of their applications as adsorbents, catalysts and separation media for advanced materials applications. Until the late1980's, most mesoporous materials were amorphous and often had broad pore size distributions. In the early 1990s, it was the emergence of a well ordered mesoporous materials[3], but recently this area extended to many metal oxide systems other than silica, and to the novel organic-inorganic hybrid mesoporous materials [4].These new silicate materials possess extremely high surface areas and narrow pore size distributions. Rather than an individual molecular directing agent participating in the ordering of the reagents forming the porous materials. This supramolecular directing concept has led to a family of materials whose structure, composition, and pore size can be tailored during synthesis by variation of the reactant stiochiometry, the nature of the surfactant molecule, the auxiliary chemicals , the reaction conditions, or by post-synthesis functionalization techniques. Fig.1.2 shows the different structures of the M41S family [5].

Fig. 1.2 The M41S materials, MCM-50 (layered), MCM-41 (hexagonal)

The formation of these materials is based on the concept of a structural directing agent or template. Templating has been defined as a process in which an organic species functions as a central structure about which oxide moieties organize into a crystalline lattice [6, 7].So that upon the removal of the templating structure, its geometric and electronic characteristics are replicated by the inorganic materials [8]. The definition has also been elaborated to include the role of the organic molecules such as: (a) space-filling species; (b) structural directing agents; and (c) templates [6]. In M41S materials, a liquid crystal templating (LCT) mechanism (Fig 1.3)was proposed by the Mobil scientists in which supramolecularas assemblies of surfactant micelles (e.g., alkyl trimethylammoniumsurfactants) act as structure rectors for the formationof the mesophase . This mechanism behind the composite mesophase formation is best understood for the synthesis under high pH conditions. Under these conditions, anionic silicate species, and cationic or neutral surfactant molecules, cooperatively organize to form hexagonal, lamellar, or cubic structures. In other words, there is an intimate relationship between the symmetry of the mesophases and the final products [9]. The composite hexagonal mesophase is suggested to be formed by condensation of silicate species formation of a(sol-gel) around a preformed hexagonal surfactant array or by adsorption of silicate species onto the external surfaces of randomly ordered rod-like micelles through columbic or other types of interactions. Next these randomly ordered composite species spontaneously pack into a highly ordered mesoporouse phase with an energetically favorable hexagonal arrangement, accompanied by silicate condensation. This process initiates the hexagonal ordering in both the surfactant template molecules and the final product [10] .

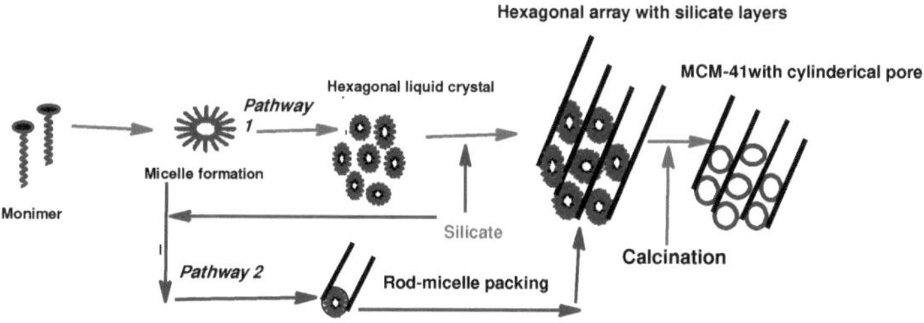

Fig. 1.3 Liquid crystal templating mechanism via two possible path ways

Other researchers further revised this liquid crystal templating mechanism [11]. The mechanism was Studied by 14N NMR spectroscopy. They concluded that the randomly ordered rod-like organic micelles interact with silica species to form two or three monolayers of silica on the outer surfaces of the micelles. Then these composite species spontaneously self-organize into a long range ordered structure to form the final hexagonal packing mesoporous. Moreover, they indicated that in the case of tetraethylorthosilicate as silica source, the concentration of the surfactant should be equal to or higher than the critical micelle concentration in order to obtain hexagonal mesopourous materials. In addition to the previously proposed mechanism, there are two other suggested liquid-crystal template mechanisms. The first mechanism was put forward by Monnier et al.[12]. It was proposed that the surfactant is initially present in the lamellar phase regardless of the final product. This lamellar mesophase transforms to the hexagonal phase as the silicate network condenses and grows. The second mechanism was proposed by Steel et al.[30]. They suggested that, as the silicate source is introduced into the reaction gel, it dissolves into the aqueous regions around the surfactant molecules, and then promotes the organization of the hexagonal mesophase. The silicate first becomes ordered into layers between which the hexagonal mesophases of micelles are sandwiched. Further ordering of the silicate results in the layers wrinkling, closing together, and growing into hexagonal channels.

1.2 Surfactants

Surfactants are molecules that have tendency to adsorb at the surfaces and interfaces. One part of the molecule is hydrophilic(head) and the other is hydrophobic(tail).Surfactant with amphilic character adsorb on the interface to reduce the free energy of phase boundary. With increasing concentration surfactant molecules combine together to form micelles to decrease the system entropy [14,15,16]. Below the initial concentration the monoatomic molecules aggregate to form isotropic micelles, which called critical micelle concentration (CMC). One popular way to classify the surfactants is according to their charge in the polar head:

1] Anionic: this family of surfactants has negatively charged polar head groups. The non-polar group is used to be a large hydrocarbon chain between C_{12} and C_{18} range. The polar groups mostly used in this kind of surfactants are carboxylates sulphates, sulfonates and phosphates.

2] Cationic: they have a positive charged polar head group and a large alkyl chain as anon-polar group. This family is based on nitrogen atom (fig.1.4). Amine and quaternary ammonium-based products are common as a head group.

Fig. 1.4 Cetyltrimethylammonium Bromide (CTAB)

3] Non-Ionics: non-ionic surfactants have a polyether or a polyhydroxyl unit as the polar group. Actually the poly(ethylene oxide) is the most common polar group. As a non-polar group poly(propylene oxide) is probably the most common. In this family, we should emphasize on the block copolymers. They are composed of blocks of different polymerized monomers triblock copolymer, like:

- PI04 (PEO_{27}-PPO_{61}-PEO_{27}).
- PI 23 (PEO_{20}-PPO_{70}-PEO_{20}).

4] Zwitterions: they have two different charges of different signs on their head group giving a neutral charge. The most common positive charge is given by an ammonium group, the source of negative charge may vary, but carboxylate is by far the most common.

1.3   Chemistry of surfactant/silicate solutions

The structural formula of mesoporous materials is based on the surfactant molecules. As the concentration process continues hexagonal close packed arrays appear, producing the hexagonal phases [17]. Then coalescence of the adjacent, mutually parallel cylinders to produce the lamellar phase. In some cases, the cubic phase also appears prior to the lamellar phase. The cubic phase is generally believed to consist of complex, interwoven networks of rod-shaped aggregates [18,19].There is many condition affect the formation of a particular phase in a surfactant aqueous solution like the concentrations ,the nature of the surfactant itself, such as the length of the hydrophobic carbon chain, hydrophilic head group, and counter ion in the case of ionic surfactants, environmental parameters, such as pH, temperature, ionic strength, solvent, and other additives. It is important to note that a high surfactant concentration, high pH, low temperature , and low degree of silicate polymerization support the formation of cylindrical micelles as well as the hexagonal mesophases [14]. The possible types of interactions between the organic and the inorganic parts that drive the formation of the mesophases depend on the charge on the surfactant S^+

or S^- are cationic and anionic surfactants respectively , the inorganic species (I^+ or I^-), and on the presence of mediating ions (X^- or M^+). All permutations enabling columbic attraction are possible, *i.e.*, S^+I^-, S^-I^+, $S^+X^-I^+$ or $S^-M^+I^-$.Subsequently, three other pathways were also discovered. Neutral (S^o) or nonionic (N^o) species can interact with uncharged inorganic species by hydrogen-bonding (S^oI^o or N^oI^o). Molecules with a covalent bond between the surfactant and inorganic parts were directly assembled (S-I). Fig.1.5 shows the interactions at the interface between the surfactant (S,I) and the silicate species.

(a–d) ionic interactions; (e) and (f) hydrogen bonding; (g) covalent bond

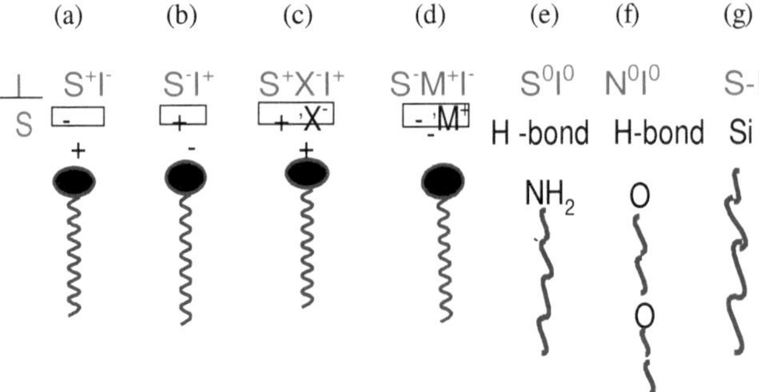

Fig. 1.5 Synthesis routes to mesoporous materials with the emphasis on silicates.

The interactions between the inorganic species and the head groups of the surfactants with consideration of the possible synthetic pathway in acidic, basic, or neutral media is discussed in table 1.1, firstly S^+I^-, $S^+X^-I^+$, $S^-M^+I^-$, S^-I^+ are electrostatic attraction ,but S^oI^o/N^oI^o are hydrogen bonds .

Table 1.1 Interactions between the inorganic species and the head groups of the surfactants.

Route	Interactions	Symbols	conditions	Classical products
S^+I^-	Electrostatic Coulomb force	S^+ cationic surfactants, I^- ,anionic silicate species	basic	MCM-41,MCM-48,MCM-50,SBA-6,SBA-2.SBA-8,FDU-2FDU-11,FDU-13
S^-I^+	electrostatic Coulomb force	S^- anionic surfactants, I^+ ,transitionmetal ions ,as Al^{3+}	A aqueous	Mesoporous alumina .
$S^+X^-I^+$	Coulomb force double layer hydrogen bond	S^+ cationic surfactants, I^+ ,cationic silicate species, $X^- = Cl^-,Br^-,I^-, SO_4^{2-}$	Acidic	SBA-1,SBA-2,SBA-3
S^0H^+ X^-I^+	Coulomb force double layer hydrogen bond	S^0 nonionic surfactants, I^+ cationic silicate species, $X^- = Cl^-,Br^-,I^-, SO_4^{2-}$	Acidic pH~2	SBA-n(n=11,12,15,16)FDU-n(n=a,5,12)
S^0I^0 N^0I^0	Hydrogen bond	S^0, nonionic surfactants, oligomeric alkyl PEO surfactants, and triblock copolymers; N^0, organic amines, CnH_2n+1NH_2, $H_2NCnH_2n+1NH_2$; I^0, silicate species, aluminate species	neutral	HMS, MSU, disordered worm-like mesoporous silicates

1.4 Sol-gel method

The sol-gel technique has been in practice since the 1930s, It described as:" Formation of an oxide network through poly condensation reactions". A sol is a stable dispersion of colloidal particles ($\geq$ 200 nm) in a solvent. An aerosol is particles in a gas phase; a gel network is built from agglomeration of colloidal particle which encloses a liquid phase. A sol-gel process occurring in several steps: Hydrolysis, condensation, ageing and drying.

1.4.1 Reactions in acidic environments

The oxygen atom in Si-OH or Si-OR is protonated and H-OH or H-OR are good leaving groups. The electron density are shifted from the Si atom, making it more accessible for reaction with water (hydrolysis or silanol condensation).

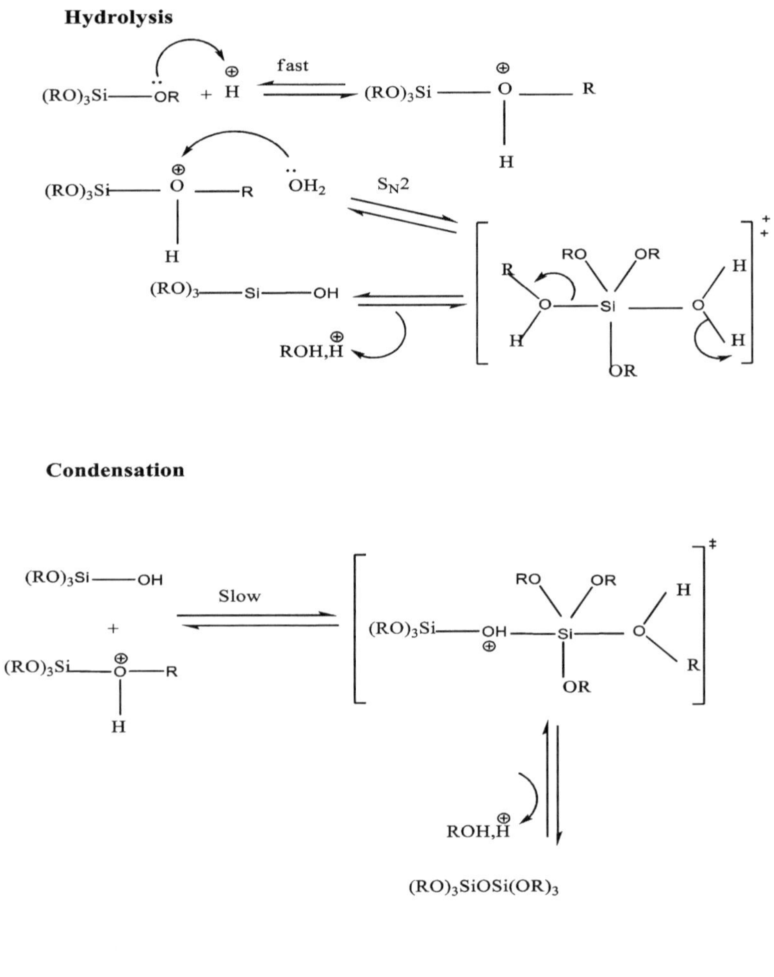

Scheme 1.1 Reaction schemes for the acid-catalyzed hydrolysis and condensation of a silicon alkoxide precursor.

1.4.2 Reactions in basic environments

Nucleophilic attack by OH- or Si-O- on the central Si atom. These species are formed by dissociation of water or Si-OH The reactions are of SN2 type where OH- replaces OR- (hydrolysis) or silanolate replaces OH- or OR- (condensation).

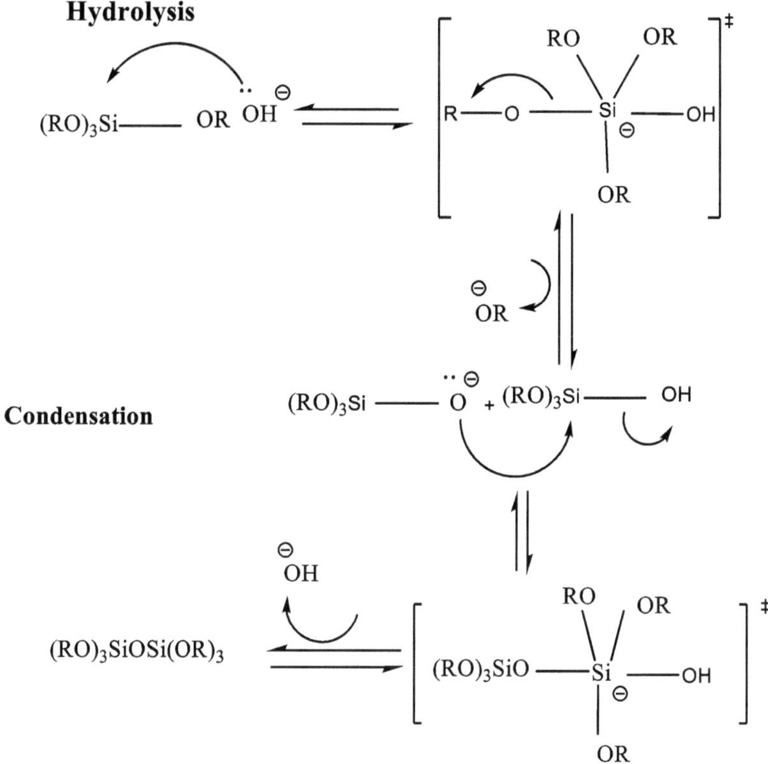

Scheme 1.2 Reaction schemes for the base-catalyzed hydrolysis and condensation of a silicon alkoxide precursor.

1.4.3 Condensation

As condensation reactions progress, the sol will set into a rigid gel. Since the reactions occur within a liquid alcoholic solvent, condensation reactions result in a three-dimensional oxide network $[M-O-M]n$

1.4.4 Aging

As the viscosity rapidly increases the solvent is "trapped" inside the gel. The structure may change considerably with time, depending on pH, temperature and solvent.

1.4.5 Drying

When the liquid is removed from the gel, an aerogel is formed and structure is maintained but a xerogel is formed if the structure is collapsed [20].

1.5 SBA-15 mesoporous silica

SBA-15 (Santa Barbara Amorphous) mesoporous silica materials, is extensive area for many research, such as adsorption, chromatography , separation, catalysis, drug delivery, sensors, photonics, and gas storages, because of it has large specific surface areas $(500-1000 m^2 g^{-1})$, pore sizes (from 2 to30 nm), high hydro thermal stability,thick framework walls, small crystallite size of primary particles and complementary textural porosity . Furthermore, one of the most important properties is, the possibility to modify their chemical surface, with different organic functional groups and high surface-to-volume ratio[21]. It has hexagonal pores in a 2D array with long 1Dchannels(*p6mm* plane group)[22].

1.5.1 Acidic synthesis

Up to now, most efficient synthesis, of ordered mesoporous silica structures were carried out in acidic media. In 1994, a breakthrough for the synthesis of highly ordered mesoporous silicates under strongly acidic conditions was realized[23]. Mesoporous silicas SBA-1 were prepared by using, cetyltriethylammonium bromide as a template and TEOS as silica source. Thereafter, SBA-15 and SBA-16 large-pore mesoporous silicas were synthesized, by using triblock copolymers as templates, under acidic conditions[23]. The characteristics for an acidic synthesis are as follows:

1) pH dependence : the synthesis of mesoporous silica, is accelerated by lowering the pH values of the solution. A high acid concentration leads to a fast

precipitation rate. On the other hand, acid catalyst at low concentration favors a slow condensation rate of silicate species is better. When strong acid HCl serves as a catalyst, higher concentration of H^+ turn the block copolymer to be more hydrophilic. This is because the PEO moieties of block copolymers are readily protonized.

2) The acidic synthesis is suitable for the formation of mesoporous silicates with diverse morphologies, such as "single crystals", thin films fibers, spheres, etc[24]. It may be related to the sol-gel chemistry of silicates. Linear silicate oligomers are the main products from the hydrolysis of silicates under acidic conditions that favor various regular morphologies. Base catalysis leads to a fast polymerization and condensation of silicates, yielding 3D silicate networks. The morphology is sometimes difficult to control. Spherical particles are the most.

3) Irreversible reaction: the irreversible polymerization of silicate species will lead to failure of the synthesis once the gel forms. In contrast, the hydrolysis of silicates is reversible under basic conditions. Ordered mesostructures can be synthesized even if a gel appears.

4) Simple silica source : siliceous oligomers and monomers are suitable precursors owing to the irreversible polymerization of silicates under acidic conditions. TEOS is the optimal choice.

5) Low processing temperature : the synthesis of mesoporous silica (SBA-3) is carried out at room temperature by using cationic surfactant. Heating or hydrothermal treatment is not adopted[24].

1.5.2 Basic synthesis

Under basic conditions, in the pH range from 9.5 to 12.5 the polymerization and cross-linkage of silicate species are reversible. Therefore, silicate precursors that can be used to prepare ordered mesoporous Silicas are diverse, for example silica gels, colloidal sols, silica aerogels. Hydrothermal treatment is necessary to prepare ordered mesoporous silicates when they are used as silicate precursors. Mixed silicate precursors were used in the synthesis of MCM-41 [25]. It is found that TEOS is the most convenient and efficient silicate precursor in the laboratory. Sodium hydroxide, potassium hydroxide, ammonium hydroxide can be used as a base. Therefore, high-quality MCM-41 can be synthesized in the pH range between 11.0 and 11.5.

1.5.3 Hydrothermal method

Mesoporous silicates are generally prepared under "hydrothermal conditions. The typical sol-gel process is involved in the "hydrothermal" process. A general procedure includes several steps. First, a homogeneous solution is obtained by dissolving the surfactant(s) in a solvent. Water is the most common solvent and silicate precursors are then added into the solution where they undergo hydrolysis catalyzed by an acid or base catalyst and transform to a sol of silicate oligomers. As a result of the interaction between oligomers and surfactant micelles cooperative assembly and aggregation give precipitation from a gel. During this step, micro phase separation, and continuous condensation of silicate oligomers occur. The formation of mesoporous silicates is rapid, only 3-5 min in cationic surfactant solutions[26]. The formation of meso structures is slow if nonionic surfactants are used as templates, normally in 30 min or even longer. Tetramethoxysilane (TMOS) as an inorganic silicate precursor results in a faster formation mesoporous silica structures than TEOS. In comparison with them, this can be attributed to their hydrolysis rates. Subsequent solidification and reorganization further proceed to form an ordered mesostructure. Hydrothermal treatment is then carried out to induce the complete condensation and solidification and improve the organization. The resultant product is cooled down to room temperature filtered washed, and dried. Mesoporous material is finally obtained after the removal of organic template(s). Neutral solutions are unsuitable to get ordered silicate mesostructures, because of too rapid polymerization and cross-linking rates of silicates at pH 6.0-8.5 to control the surfactant-templating assembly [27].

1.6 Stöber method (silica microphase)

In 1968,WernerStöber published method of creating spherical , mono-dispersed silica nanoparticles ranging in size from 0.05-2 μm [28].The diameter of silica particles from the Stober process is controlled by the relative contribution from nucleation and growth processes. The hydrolysis and condensation reactions provide precursor species and the necessary supper saturation for the formation of particles. During the hydrolysis reaction, the ethoxy group of TEOS reacts with the water molecule to form the intermediate [Si (OC_2H_5) $_4$-X (OH)X] with hydroxyl group substituting ethoxy groups. Ammonia works as a basic catalyst to this reaction; the hydrolysis reaction is initiated by the attacks of hydroxyl anions on TEOS molecules., the condensation reaction occurs immediately. Where the hydroxyl group of intermediate[Si (OC$_2$H$_5$) $_4$ -X (OH)X] reacts with either the ethoxy group of other TEOS "alcohol condensation" or the hydroxyl group of another hydrolysis intermediate "water condensation" to form Si-O-Si bridges. Mono-dispersed submicron silica particles applied as packing material for capillary chromatography[28].

1.7   Functionalized mesoporous silica

Combination the properties of organic and inorganic blocks within a single material is attractive for materials scientists because of the combination of functional groupes of organic chemistry with a thermally stable and robust inorganic substrate. This is particularly applicable to heterogeneous catalysis. Equally interesting is modification with organic functionalities such as C-C multiple bonds, alcohols, thiols, sulfonic ,carboxylic acids and amines, etc.

1.7.1 Post synthetic functionalization of silicas (Grafting)

Grafting refers to the subsequent modification of the inner surfaces of mesostructured silica phases with organic groups. This process is carried out primarily by reaction of organosilanes of the type (R'O)$_3$SiR, or less frequently chlorosilanes ClSiR$_3$ or silazanes HN(SiR$_3$)$_3$, with the free silanol groups of the pore surfaces. In principle functionalization with a variety of organic groups can be realized in this way by variation of the organic residue. This method of modification has the advantage that, under the synthetic conditions used, the mesostructure of the starting silica phase is usually retained, whereas the lining of the walls is accompanied by a reduction in the porosity of the hybrid material (albeit depending upon the size of the organic residue and the degree of occupation). If the organosilanes react preferentially at the pore openings during the initial stages of the synthetic process, the diffusion of further molecules into the center of the pores can be impaired which can in turn lead to a non homogeneous distribution of the organic groups within the pores and a lower degree of occupation. In extreme cases (e.g., with very bulky grafting species), this can lead to complete closure of the pores (pore blocking). The process of grafting is frequently erroneously called immobilization, so it use to remove a toxic environmentally relevant contaminants by adsorption[29].

1.8 Metal oxide nanoparticles

Nanoparticles usually ranging in dimension from 1-100 nanometers (nm) have properties unique from their bulk equivalent. With the decrease in the dimensions of the materials to the atomic level, their properties change. The nanoparticles possess unique physico-chemical, optical and biological properties which can be manipulated suitably for desired applications [30]. There is a number of specific methods to prepare metal oxides nanoparticles

(1) Co-precipitation method : this involves dissolving a salt precursor (chloride, nitrate, etc.) in water (or other solvent) to precipitate the oxo-hydroxide form with the help of a base. Very often, control of size and chemical homogeneity in the case of mixed-metal oxides are difficult to achieve. However, the use of surfactants, sonochemical methods, and high-gravity reactive precipitation appear as novel and viable alternatives to optimize the resulting solid morphological characteristics [31,32].

(2) Sol-gel method . In this method metal oxides are obtained via hydrolysis of precursors, usually alcoxides in alcoholic solution, resulting in the corresponding oxo-hydroxide. Condensation of molecules by giving off water leads to the formation of a network of the metal hydroxide: Hydroxyl-species undergo polymerization by condensation and form a dense porous gel. Appropriate drying and calcinations lead to ultrafine porous oxides[33].

(3) Microemulsion method , or (direct/inverse micelles) represents an approach based on the formation of micro/nano-reaction vessels under a ternary mixture containing water, a surfactant and oil. Metal precursors on water will proceed precipitation as oxo-hydroxides within the aqueous droplets, typically leading to monodispersed materials with size limited by the surfactant-hydroxide contact[34].

(4) Solvothermal methods, in this case, metal complexes are decomposed thermically either by boiling in an inert atmosphere or using an autoclave with the help of pressure. A suitable surfactant agent is usually added to the reaction media to control particle size growth and limit agglomeration.

(5) Template/Surface derivative methods, template techniques are common to some of the previous mentioned methods and use two types of tools; soft-templates (surfactants) and hard-templates (porous solids as carbon or silica). Template- and surface-mediated nanoparticles precursors have been used to synthesize self-assembly systems[31]. Gas-solid transformation methods with broad use in the context of ultrafine oxide powder synthesis are restricted to chemical vapor deposition (CVD) and pulsed laser deposition (PLD).

(6) There are a number of CVD processes used for the formation of nanoparticles among which we can highlight the classical (thermally activated/pydrolytic), metalorganic, plasma-assisted, and photo CVD methodologies[35].The advantages of this methodology consist of producing uniform, pore and reproduce nanoparticles and films although requires a careful initial setting up of the experimental parameters.

(7) Multiple-pulsed laser deposition heats a target sample (4000 K) and leads to instantaneous evaporation, ionization, and decomposition, with subsequent mixing of desired atoms. The gaseous entities formed absorb radiation energy from subsequent pulses and acquire kinetic energy perpendicularly to the target to be deposited in substrate generally heated to allow crystalline growth[36].

Metal oxide nanoparticles have wide application like zinc oxide (ZnO), it have a wide band gap semiconductor with optical, piezoelectric, and dielectric properties that render it useful for a variety of applications, ranging from transistors,[37] light emitting diodes [38] sensors [39] photo catalysis [40] and ultraviolet filters [41].Copper oxide nanoparticle(CuO) a semiconducting compound with a narrow band gap so it is used for photoconductive and photo thermal applications [42] and used as nano fluids heat transfer application. Magnesium oxide (MgO) have extensive porous structure with considerable pore volume [43]. Their high surface area and enhanced surface reactivity make it a good catalyst for many reaction [44] .

1.9 Supported metal oxide nanoparticles

When the size of metal oxide NPs decreases, their relative surface area becomes greater and their activity often increases. The use of stabilizing agent is important to stabilize the surface, minimized the leaching, prevent the aggregation of nanoparticles and minimize atom/ion leaching from the particles[45].Zeolite, alumina, graphen, carbone nanotube, organic polymers and mesoporous silica were used. There is different methods to protect metal oxide nanoparticles e.g. silica like coating, impregnation and co-condensation.

1.9.1 Impregnation

Impregnation is the most conventional method and can be easily carried out by dripping a metal precursor solution (metal acetate, chloride and nitrate) onto the mesoporous support directly then it flow by drying and calcination. With this method, the loading amount of the metal oxide can be very high, but it yields an uncontrolled growth of metal particles that occurs both inside and outside of the channels of the mesoporous silica, as well as resulting in the agglomeration of particles and a decrease in the catalytic efficiency. The most apparent advantage of the impregnation method is good preservation of the mesostructured after modification. However, it has several shortcomings. First, attachment of metal oxide on the pore surface reduced pore size and pore volume, second a limited loading level of the metal oxide because of the limited density of the reactive surface silanols [46]. Third, it's a time-consuming process as it involves two steps pre-synthesis of a parent mesostructure, followed by post-treatment with metal precursor [47-48]. Fourth, control over the loading level and uniformity of metal oxide distribution difficult to achieve [48].

1.9.2 Co-condensation (direct synthesis)

It is possible to prepare metal oxide nanoparticles supported mesostructured silica phases by the co-condensation of tetraalkoxysilanes $(RO)_4Si$ (TEOS or TMOS)] with terminal trialkoxyorganosilanes of the type $(R'O)_3SiR$ in the presence of structure-directing agents and metal precursor (metal acetate ,nitrate and chloride) flowed by drying and calcination. This method has advantages like, no pore blocking because of the homogeneous distribution of metal oxide, a higher and more uniform surface area and a better control over the surface properties of the resultant materials [49], it can lower the synthesis temperature, the pore sizes of directly synthesized samples are a bit larger than parent SBA-15. This results from addition of inorganic salts that lower the thermodynamic radius of micelles, the silica wall of the final product is ultimately thinner and the pore size slightly larger . However, the co-condensation method also has a number of disadvantages in general, the degree of mesoscopic order of the products decreases with increasing concentration of metal precursor in the reaction mixture, which ultimately leads to totally disordered products. The tendency towards homocondensation reactions, which is caused by the different hydrolysis and condensation rates of the structurally different precursors, is a constant problem in co-condensation because the homogeneous distribution of different metal oxide in the framework cannot be guaranteed. Moreover, an increase in loading of the incorporated metal oxide can lead to a reduction in the pore diameter, pore volume, and specific surface area.

1.10 LITERATURE REVIEW

1.10.1 Supporting of mesoporous silica

Metal- metal oxides introduce in mesoporous silica, are a class of important hybrid materials, that have many significant applications in a wide variety of technical fields, including adsorption, separation, optoelectronics, biological medicine and, catalysis [50].

Impregnation, co-condensation and two solvent methods have been developed for introducing metal or metal oxide into mesoporous silica [51]. In impregnation method, the loading amount of the metal/metal oxide can be very high, in order to improve the quality of the product, an ultrasonic-assisted technique has been developed to synthesize mesoporous materials [52-54].

Metal oxides such as MoO_3 have been impregnated into MCM-41with the ultrasound assistance [54]. Ru/SBA-15 has been synthesized by using the ultrasonic method [55]. It was reported that the ultrasound technique provides a unique advantage in the preparation of a highly-loaded and well-dispersed mesoporous composite[56]. It was developed a novel ultrasonic post-grafting method was developed to obtain highly-loaded and well-dispersed CuO nanoclusters inside the channels of mesoporous silica (SBA-15). It was used for hydroxylation of benzene to phenol. Trimethylchlorosilane (TMCS) and aminopropyltriethoxy silane (APTES) were used as modification surfactants, and copper nitrate solution was used as the precursor., they found CuO clusters were well-dispersed in very small sizes. [57].

Novel porous materials with mesoporous structure and strong basicity by dispersing MgO on SBA-15 in three different methods was successfully prepared, including impregnation microwave irradiation and their combination, through magnesium acetate path. The efficiency of microwave irradiation was proven to be very high, but it fails to directly disperse magnesia itself on SBA-15, and the lack of ions in surface of carrier was excluded to account for this failure[5 8].

Iron oxide were supported on ordered mesoporous SBA-15 by impregnating as-synthesized SBA-15 with a methanolic solution of $Fe(NO_3)_3$. $9H_2O$, characterization revealed that iron oxide was present as highly dispersed nanoclusters in the well-ordered mesoporous channels of SBA-15. The supported material still maintained its ordered mesoporous structure similar to SBA-15, possessed high surface area, large pore volume, and uniform pore size. The benzylation of benzene by benzyl chloride showed that iron oxide nanoclusters-supported SBA-15 was a very active catalyst and able to activate the reactant at relatively low temperature such as 313 K.   Moreover, the catalyst could be reused [59].

Post grafting technique was used to incorporate various amounts of tin using two different metal precursors, tin acetate and tin chloride. Higher amounts of tin could be incorporated into SBA-15 using tin acetate. Adding dilute HCl to parent SBA-15 in the acetate precursor could have increased the number of silanol groups, thereby facilitating the incorporation of higher amounts of tin. XRD data indicate a good mesoscopic order. The characteristic hexagonal features of SBA-15 were maintained in Sn-SBA-15 samples. Incorporation of tin does not affect the original pore structure of the parent SBA-15 even at high tin loading. The silanol groups on the internal walls of SBA-15 are suggested to be the sites for tin incorporation [60].

A series of yttrium (Y)-containing mesoporous SBA-15 were prepared using a sole gel method with various Y/Si molar ratios was investigated as the supporting material of nickel (Ni) catalysts for the methane reforming with CO_2. The highly ordered hexagonal structure of SBA-15 was well-retained after the incorporation of yttrium at the molar ratio of Y/Si (0.04). The presence of yttrium in the framework of SBA-15 in Ni catalysts effectively enhanced the formation of the Ni metallic particles with small size [61].

Simple and effective method. Impregnation method were developed for synthesis of Pd-doped mesoporous silica SBA-15(Pd/SBA-15), due to the large surface area and high catalytic behavior of Pd/SBA-15, carbon paste electrode modified with Pd/SBA-15 (Pd/SBA-15/CPE) was prepared. Pd/SBA-15/CPE was used for oxalic acid (OA) detection in real samples. The proposed method showed a good result, indicating that the present modified electrode could be applied to determine OA in food samples [62].

In another hand, vanadium oxide were grafted on mesoporous silica SBA-15, using a controlled grafting process, the spectroscopic results revealed that under dehydrated conditions, the grafted vanadium domains are highly dispersed on the

SBA-15 surface, composed predominately of isolated VO$_4$ units with distorted tetrahedral coordination. Methanol oxidation was used as a chemical probe reaction, to examine the catalytic properties of these catalysts. At low vanadium loading, the vanadium species grafted on the surface show structural properties similar to those of vanadium-incorporated MCM-41 catalyst. However, the present mesoporous V-SBA-15 catalysts in the oxidation of methanol to formaldehyde, show remarkable catalytic performance compared with that of VOx/SBA-15 catalysts synthesized through a conventional wet impregnation method, which has been attributed to the homogeneous dispersion, and uniformity of the catalytic vanadium species achieved on the SBA-15 support with large pore diameter and surface area [63].

SBA-15 mesoporous silica was modified with metal (Al, Ti, Cu, Fe) oxides by the molecular designed dispersion (MDD) method using acetylacetonate complexes of metals as precursors of the catalytically active components. Species on the SBA-15 surface significantly increased its acidity, mainly by generation of strong Lewis acid sites. Copper and iron deposited on the surface of pure SBA-15 were present nearly exclusively in the form of mononuclear cations. Deposition of Fe or Cu on the SBA-15 supports modified with aluminum or titanium resulted in a formation of significant amounts of oligomeric metal oxide clusters. The modification of the silica surface with titanium or aluminum prior to the deposition of iron or copper significantly improved the activity of the SBA-15 based catalysts[64,65].

A simple solvothermal impregnation method was used to prepare ZnO nanoparticles supported on MCM-41 and SBA-15. The influence of the ZnO loading of different supports on the structural characteristics and the photocatalytic activity toward degradation of methylene blue in water under ultraviolet irradiation were investigated. Much smaller influence of impregnation with ethanolic zinc salt solution on the porosity was observed for SBA-15compared with MCM-41. Finally, the adsorption and photo catalytic activity of the ZnO/mesoporous materials depend on porous characteristics of the support materials[66].

Pt nanoparticles were dispersed in the mesochannels of SBA-15 silica by a simple one-pot, co-assembly method. Metallic Pt can be encapsulated in the mesochannels by changing the amount of Pt ions in the starting materials. The catalytic performances of Pt–SBA-15 were evaluated for methylcyclohexane dehydrogenation. Compared with conventional supported Pt–SiO$_2$ catalysts, the Pt–SBA-15 catalysts show higher stability. TEM studies show that the confinement effect of ordered mesochannels of SBA-15 may restrict the further growth of Pt nanoparticles during the reaction [67].

1.10.2 Amine functionalization of mesoporous silica

Functionalization /modification of mesoporous materials has played a vital role in various technological aspects such as improvement of thermal/hydrothermal stability, immobilization of enzymes, development of new catalyst and adsorbent, and nanotechnology [68,69,70]. Generally, there are two approaches to surface functionalization of mesoporous silica materials, i.e. grafting (also known as post-synthesis) and direct synthesis or co-condensation [69,70].

Grafting method (post-synthesis) was used to functionalized MCM-41 mesopourous silica by 2-(3-(2-aminoethylthio)propylthio)ethanamine the resulting material adsorbed Hg(11) ions from aqeuos solution and showed abetter ability for Hg(11) ions adsorption [71].

The NH$_2$ -groups were successfully grafted onto the SBA-15 mesopourous silica surface. The analysis method showed the successful incorporation of the-NH$_2$ onto the surface .Its revealed that grafting is an effective method to obtain a high loading of NH$_2$-groups onto the SBA-15 with high adsorption rates 99.9%, 99.7% , 99.8%, 99.5 % for Pd^{2+},Cr^{3+},Cd^{2+}, Ag$^+$, respectively [72] .

Functionalization by co-condensation method was carried out by co-condensation of TEOS , APTES, mercaptopropyl-trimethoxysilane (MPTMS), phenyltrimethoxysilane (PTMS), vinyltriethoxysilane (VTES), and 4-(trieth-oxysilyl) butyronitrile (TSBN) in the presence of nonionic triblock co-polymer Pluronic P123 under acidic conditions. It was found that functionalization of large-pore mesoporous silicas by direct synthesis is effective and controllable. It offers a higher and more uniform surface coverage of functional groups and a better control over the surface properties of the resultant materials, with multifunctions for various uses [73].

Functionalized SBA-15 mesoporous silica materials have been synthesized through a simple co-condensation approach of bis-[3-(trimethoxyosily)propyl] amine (BTPA) and tetraethyl orthosilicate (TEOS).Resulting materials showed high adsorption capacities for Hg(11) ions in acid solutions[74].

1.11 Aim of the present work:

- This study aimed to:

- Synthesis mesoporous silica SBA-15 by hydrothermal process in acidic media.

- Synthesis mesoporous silica SBA-15 supporting metal oxides nanoparticles CuO and ZnO with different ratio (5%,10%,15%,20%) by impregnation method.

- Functionalize mesoporous silica supporting metal oxide nanoparticles with propyl amine.

- Characterize the synthesized samples using X-ray diffraction (WAXS) ,(SAXS), Fourier transform Infrared Spectrometry (FT-IR), Ultraviolet-visible spectrometers (UV-Vis), and Thermo gravimetric analysis (TGA).

- Test the extraction and removal of E124 food synthetic dye from water by functionalized mono amine SBA-15 mesoporous silica supporting ZnO (20%), and functionalized mono amine SBA- 15 mesoporous silica free ZnO (20%).

CHAPTER TWO

Experimental

2.1   Chemicals and reagents.

All chemicals purchased analytical grade and used as received without further purification. Tetraethylorthosilicate (TEOS) [($C_2H_5O)_4Si$], triblock copolymer Pluronic P123 ($EO_{20}PO_{70}EO_{20}$),Toluene($C_6H_5CH_3$),Zinc acetate dehydrate [$Zn(CH_3COO)_2.2H_2O$], Copper acetate monohydrate[$Cu(CH_3COO)_2.H_2O$] were purchased from Aldrich company. 3- Aminopropyltrimethoxysilan and hydrochloric acid HCl ,were purchased from Merck. Glassware's used in the experimental work were washed with distilled water and dried at 100 ^{0}C.

2.2   Synthesis

2.2.1 Synthesis of mesoporous silica SBA-15

Mesoporous silica SBA-15 was synthesized as described in literature[75]. When 2.0 g triblock copolymer Pluronic P123 was dissolved in30 ml HCl (2M) and 15 ml H_2O at room temperature. 4.4g (0.02mol) of TEOS (tetraethylorthosilicate) was added to the stirred solution. The resulting mixture was stirred for 24h at RT, then auto clave at 100 ^{0}C for another 48 h under static condition. The as-made sample was recovered by centrfusion, washed with distilled water and dried at 100^0C for 24 h, Finally the obtained material was calcined at 550 ^{0}C in air for 6 h to remove the copolymer surfactant scheme (2.1).

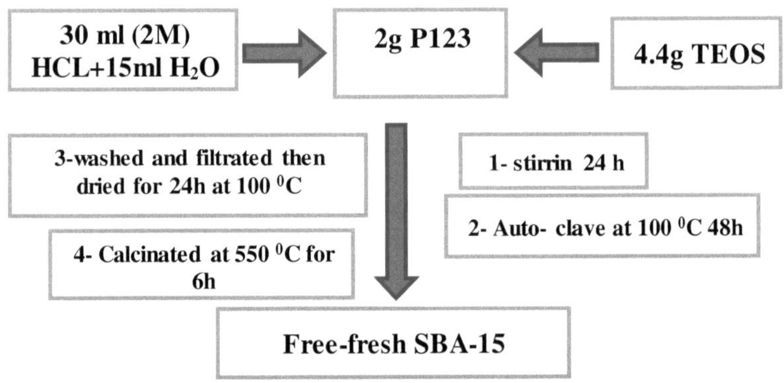

Scheme 2.1 Synthesis of mesoporous silica SBA-15

2.2.2 Preparation of metal oxides coated mesoporous SBA-15 silica

2.2.2.1 Impregnation Method

The metal oxides coated mesoporous SBA-15 Silica (ZnO/SBA-15 or CuO/ SBA-15)materials were prepared as previously reported by impregnation method [86] ,by treatment of the corresponding metal acetate solution of different concentration (5,10,15,20%) with mesoporous SBA-15 Silica. The mixture was stirred at 80 ^{0}C,then dried at 100 ^{0}C overnight. The materials were then calcinated at 550 ^{0}C for 6 hours as given in scheme (2.2).

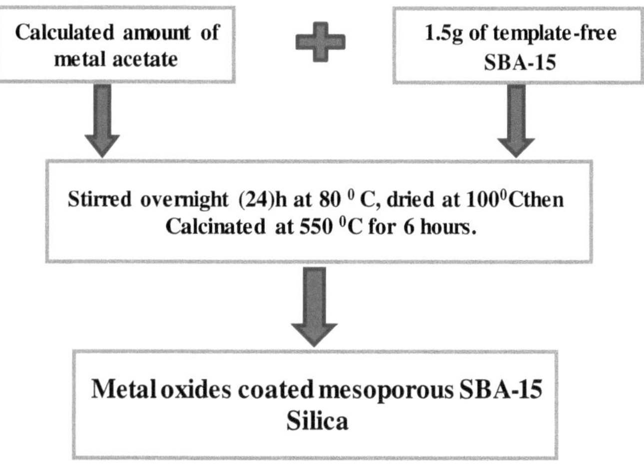

Scheme 2.2 Impregnation method

2.2.3 Monoamine functionalization of metal oxide coated SBA-15 silica

Amine-functionalized metal oxide coated SBA-15silica were prepared as previously described [77] by disperse 0.5g of(20%Metal oxide coated SBA-15 Silica) prepared by impregnation in 15ml dry toluene, followed by adding 0.075g, 0,002 mole of 3-aminopropyltrimethoxysilan coupling agent. The mixture was refluxed at 110^0 C. The amine functionalized material was filtered off washed with ethanol and dried in vacuum at 60^0 C scheme (2.3).

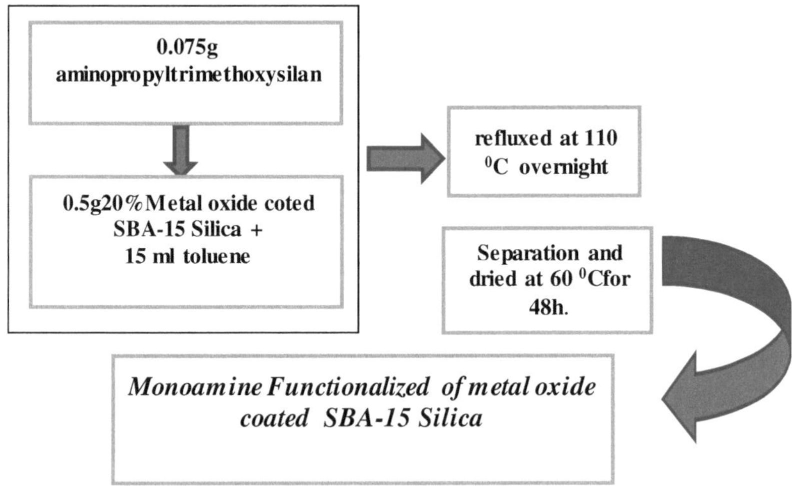

Scheme 2.3 Functionalization by monoamine

2.2.4 Free metal oxide coated SBA-15 silica

Free metal oxide coated SBA-15 silica were obtained by treating 0.5 g of metal oxide coated SBA-15 silica (20%) with(20 ml,2M) concentrated hydrochloric acid with continuous stirring for 12 h. The product were separated, washed with distilled water and dried in vacuum at 80 ^{0}C.

2.2.5 Preparation of colloidal solution

For optical absorption and photoluminescence measurements, 0.0015g of sample was dissolved in 3ml of 20% HY solution .Optical absorption measurements was taken after one day .

2.3 Adsorption study of synthetic dye

Batch method was used by shaken 0.10 g of the samples (SBA-15–NH_2,ZnO/SBA-15-NH_2,free ZnO/SBA-15-NH_2) with 5 ml of E124 (70ppm) the study was examined by UV-vis spectra .

2.4 Characterization

The following techniques and methods were used for structure characterization of SBA-15,metal oxide coated SBA-15 Silica, and their amino functionalized metal oxide coated SBA-15 Silica .

2.4.1 Fourier Transform Infra Red (FT-IR)

Fourier Transform Infrared (FTIR) spectroscopy is one of the most common spectroscopic technique used by organic and inorganic chemists. Simply ,it is the absorption measurement of different IR frequencies by a sample positioned in the path of an IR beam .The main goal of IR spectroscopic analysis is to determine the chemical functional groups in the sample. Different functional groups absorb characteristic frequencies of IR radiation. Using various sampling accessories, IR spectrometers can accept a wide range of sample types such as gases, liquids ,and solids . Thus, IR spectroscopy is an important and popular tool for structural elucidation and compound identification. FTIR spectra were recorded using Fourier Transform Infrared spectrophotometer (Frontier (Perkin Elmer); The samples are measured on a Zinc Selenide Crystal, it is working as a Multiple Reflection ATR system (Attenuated total Reflection). Fig. 2.1 [78].

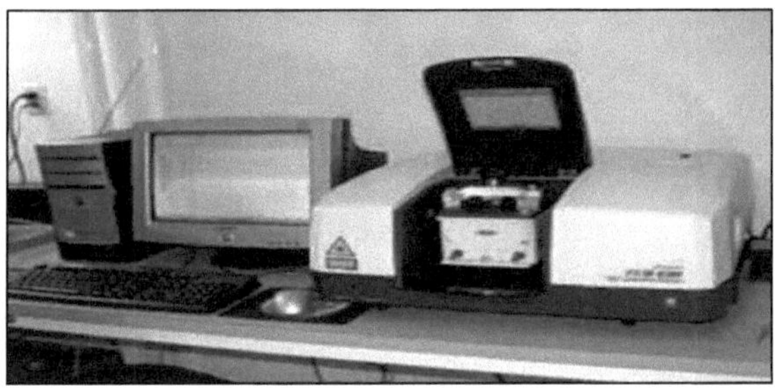

Fig. 2.1 FT-IR spectrophotometer

2.4.2 Ultraviolet-visible spectroscopy (UV-vis)

Ultraviolet-visible (UV-vis) spectroscopy or Ultraviolet-Visible (UV-Vis) spectrophotometry (UV-vis or UV/vis) refers to absorption spectroscopy in the ultraviolet-visible spectral region. This means it uses light in the visible and adjacent (near-UV and near-infrared (NIR)) ranges. The absorption in the visible rang directly affects the perceived color of the chemicals involved. In this region of the electromagnetic spectrum, molecules undergo electronic transitions. This technique is complementary to fluorescence spectroscopy, in that fluorescence deals with transition from the excited state to the ground state, while absorption measures transition from the ground state to the excited state, UV/vis spectroscopy is routinely used in the quantitative determination of solutions of transition metal ions highly conjugated organic compounds, and biological macromolecules Fig. 2.2 [79].

Ultraviolet-visible absorption spectra were recorded on a UV-vis spectrophotometer Shimadzu, UV-2400 in the wavelength range from 200 to 800 nm.

$$(\alpha h\nu)^2 = h\nu\text{-}Eg$$

α = absorbance coefficient (μm).

$(\alpha h\nu)^2$ = absorbance coefficient (μm eV).

Eg= E = Band gap energy (eV) .

hν= photon energy = 1.24 μm-eV.

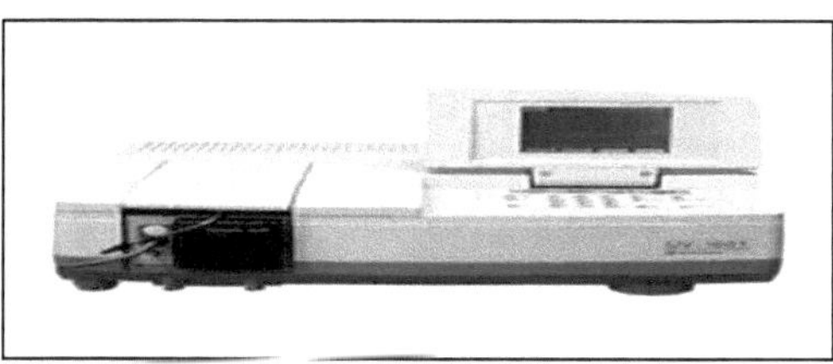

Fig. 2.2 UV-vis spectrophotometer

2.4.3 Photoluminescence spectroscopy (PL)

Photoluminescence (PL) is the spontaneous emission of light from a material under optical excitation. The excitation energy and intensity are chosen to probe different regions and excitation concentrations in the sample. PL investigations can be used to characterize a variety of material parameters. PL spectroscopy provides electrical (as opposed to mechanical) characterization, and it is a selective and extremely sensitive probe of discrete electronic states. Features of the emission spectrum can be used to identify surface, interface, and impurity levels and to gauge alloy disorder and interface roughness. The intensity of the PL signal provides information on the quality of the surfaces and interfaces. Under pulsed excitation Fig. 2.3 [80].

PL spectra were recorded with a spectrofluorometer (JASCO, FP- 6500); the excitation wavelength was selected to be 310 nm. The scan rate was set at 600 nm/min with the entrance and exit slit width of 5 nm.

Fig. 2.3 PL spectrofluorometer

2.4.4 Thermo gravimetric Analysis (TGA)

Thermal gragravimetric analysis(TGA) is a method of thermal analysis in which changes in physical and chemical properties of materials are measured as a function of increasing temperature (with constant heating rate), or as a function of time (with constant temperature and/or constant mass loss).TGA can provide information about physical phenomena, such as second-order phase transitions including vaporization, sublimation, absorption, adsorption, and desorption. Likewise, TGA can provide information about chemical phenomena including chemisorptions, desolvation (especially dehydration), decompositionin and solid-gas reactions (e.g., oxidationor reduction) .TGA is commonly used to determine selected characteristics of materials that exhibit either mass loss or gain due to decomposition ,oxidation, or loss of volatiles (such as moisture).common applications of TGA are:

1. materials characterization through analysis of characteraistics decomposition patterns.
2. studies of degradation mechanisms and reaction kinetics.
3. determination of organic content in a sample.
4. determination of inorganic (e.g.ash) content in a sample, which may be useful for corroborating predicted material structures or simply used as chemical analysis. It is an especially useful technique for the study of polymeric materials, including thermoplastics, thermosets ,elastomers ,composites ,plastic films, fibers coatings and paints Fig. 2.4 [81].

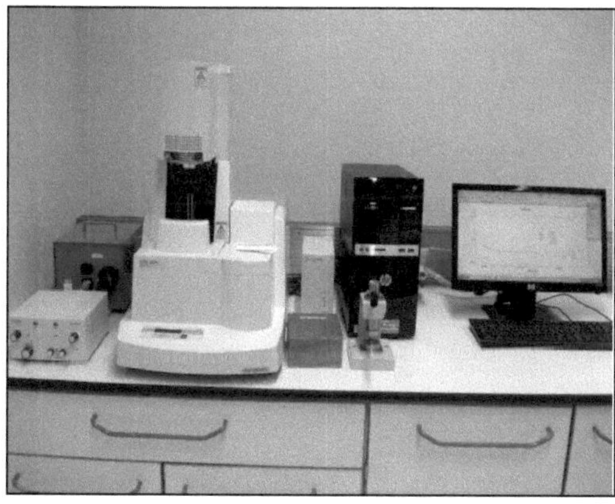

Fig. 2.4 Thermo gravimetric analyzer (TGA)

2.4.5 X-Ray Diffraction (X-RD)

The discovery of X-rays in 1895 enabled scientists to probe crystalline structure at the atomic level. X-ray diffraction has been in use in two main areas, for the finger print characterization of crystalline materials and the determination of their structure. Each crystalline solid has its unique characteristic X-ray powder pattern which may be used as"fingertanprint" for its identification. Once the material has been identified, X-ray crystallography may be used to determine its structure, i.e. how the atoms pack together in the crystalline state and what inter atomic distance and angle are etc., X-ray diffraction in one of the most important characterization tools used in solid state chemistry and materials science . We can determine the size and the shape of the unit cell for any compound most easily using X-ray diffraction. Fig. 2.5 [82].

The X-ray diffraction (XRD) patterns of the dried as-prepared and classified samples were obtained using an X-ray diffractometer PANalytical X0 pert (PANalytical) with Cu Ka radiation (0.154 nm wavelength) under 40 kV and 200 mA. The Scherrers equation was used to determine bulk crystallite diameter of a sample. The following is a general form of the Scherrers equation:

d = crystallite diameter (nm).

K = crystallite shape constant (using 0.9 for spherical shape).

λ = x-ray wavelength (0.154 nm).

B = Full-width at half max at Bragg angle of interest.

θ = Bragg Angle (angle of interest)

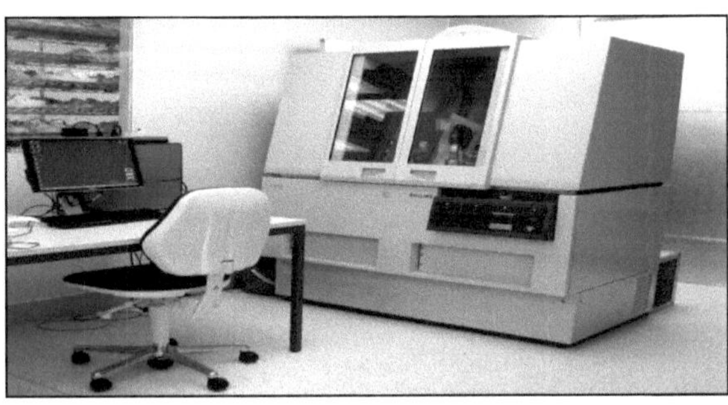

Fig. 2.5 X-ray diffractometer

2.4.6 Small angle x-ray scattering (SAXS)

SAXS became a major tool to rapidly and comprehensively characterize macromolecular and nanostructured systems Fig. 2.6. The main principles of SAXS were developed in the late 1930s by A. Guinier with his studies of metallic alloys. Already in the first monograph on SAXS by Guinier and Fournet (1955) it was demonstrated that the method yields not just information on the sizes and shapes of particles but also on the internal structure of disordered and partially ordered systems.

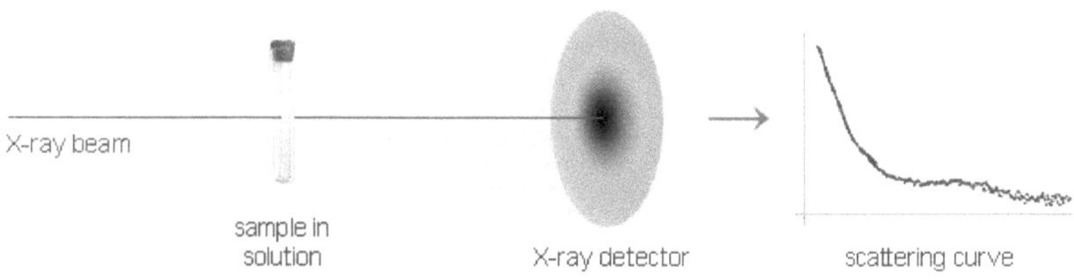

Fig. 2.6 Small angle x-ray scattering (SAXS)

Conceptually, a SAXS experiment is simple: a sample is illuminated by X-rays and the scattered radiation is registered by a detector. As the SAXS measurements are done very close to the primary beam ("small angles"), the technique profits immensely from the brilliance of X-ray photon beams provided by particle accelerators known as synchrotrons [83].

CHAPTER THREE
Results and Discussion

3.1 Synthesis

Two main strategies have been developed to incorporate metal oxides into the channels of mesoporous silica SBA-15,MCM-41 and the pores of zeolites. Impregnation method[84-87], and co-condensation method [88,89] . Generally, the first strategy seems difficult to be completed avoid adsorptions of the ZnO precursor on the outer surface of the host template. The uncontrolled ZnO aggregation on the external surface of mesoporous silica will form in subsequent calcinations. The second strategy involves complicated process and has low yield. More recently, two-solvent method may be employed to prepare a nanocomposite of ZnO clusters supported in mesoporous silica[90]. In our present work we used the impregnation method.The systematic procedure for impregnation of metal oxides (CuO or ZnO) is presented in Scheme 3.1 .SBA-15 silica was firstly prepared by using pluronic P123 triblock copolymer as previously prepared [91], see Scheme 3.1. The pluronic triblock copolymer was removed either by calcinations at 600 ^{0}C or by washing the material with hot ethanol. The second step was to introduce the metal containing precursor into the silica pores allowing the precursor to decompose inside the pore during thermal treatment, controlling the growth of the metal oxide nanoparticles as previously reported[84]. The impregnated metal oxide mesoporous SBA-15 silica are obtained, labeled as ZnO/SBA-15and CuO/SBA-15. Monoamine functionalized ZnO/SBA-15-NH$_2$and CuO/ SBA-15-NH$_2$were also obtained by interaction between the metal oxide coated SBA-15silica composites and monoamine silane agent in toluene using a previous reported method[77] . Removal of metal oxides nanoparticles from the functionalized materials were obtained using concentrated HCl. Its clear that metal oxides clusters were entrapped into the silica network pores. These materials were examined by FT-IR, TGA , SEM, TEM and X-RD techniques .

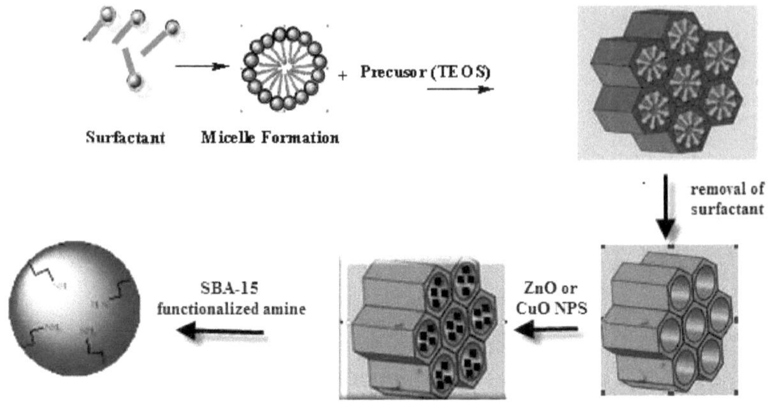

Scheme 3.1 Schematical diagram of synthesis mechanism

3.2 X-RD analysis

3.2.1 Small angle X-ray scattering (SAXS)

The small angle X-ray scattering (SAXS) patterns for SBA-15silica obtained by extraction and calcination for removal P123 surfactant which are labeled as SBA-15ext. and SBA-15cal. are presented in Fig. 3.1a & respectively. The original SBA-15 silica samples showed a typical pattern of a hexagonal phase with the occurrence of a strong peak, due to the (100) plane, and other two weak peaks, due to the (110) and (200) planes as shown in Fig.3.1a & b. The presence of three well-resolved diffraction peaks is associated with highly ordered mesoporous silica SBA-15 with a two-dimensional hexagonal symmetry (space group p6mm) [92]. There was a slight shift of all three peaks to larger angle in the case of the SBA-15-cal. silica Fig.3. 1b. This results of smaller interplanar spacing (contraction), large pore size and high surface area as confirmed by BET analysis (surface area increased from 600 m^2/g for SBA-15-ext. to 900 m^2/g for SBA-15-calc. silica). The d_{100} spacing value has changed from 9.2 for SBA-15-ext. silica to 8.8 nm for SBA-15-calc. silica, Table 3.1 .

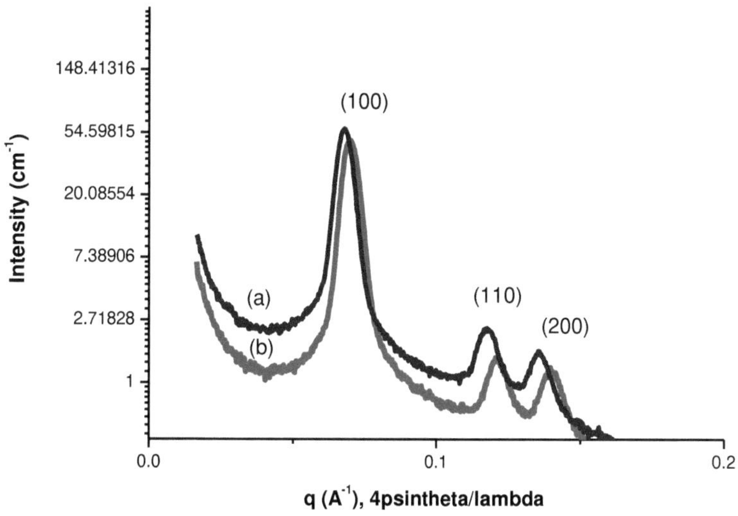

Fig. 3.1 SAXS for (a) SBA-15-ext. (b) SBA-15-cal.Silica

Table 3.1 XRD and BET result.

Materials	d_{100} (nm)	Mean Crystallite size (metal oxide) (nm)	Surface area(m^2g^{-1})	a_0(nm)
SBA-15 ext. silica	9.2	-	600	10.6
SBA-15 cal. Silica	8.8	-	900	10.1
SBA-15//CuO 20%	9.2	2.1	-	10.6
SBA-15/ZnO 20%	9.2	2.2	-	10.6

The small angle X-ray scattering (SAXS) patterns of impregnated CuO/ SBA-15 composites of different percentages (5,10,15,20%) are similar to the patterns of SBA-15 pure silica. It showed a typical pattern of a hexagonal phase with the occurrence of a strong peak, due to the (100) plane, and other two weak peaks, due to the (110) and

(200) planes as shown in Fig. 3.2. This provides evidence that the addition of inorganic precursors does not alter the mesoscopic order of SBA-15 silica. There was a shift of all three peaks to a smaller angle after impregnation process .This results of increasing the interplanar spacing d values Table3.1, this suggest that the insertion of metal oxides is probably associated with expansion of the mesoporous silica[93]. Only minor shift has been seen as the loading of copper oxide is increased from 5% to 20%.

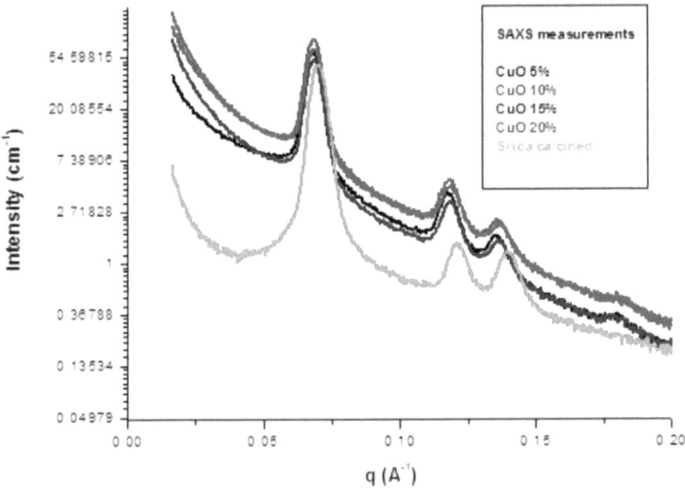

Fig. 3.2 SAXS of SBA-15 Cal. and CuO/SBA-15 composites of different percentages.

Small angle X-ray scattering patterns of impregnated ZnO/SBA-15 composites of different percentages (5,10,15 and 20%) are similar to that of SBA-15 silica. It showed a typical pattern of a hexagonal phase with the occurrence of a strong peak, due to the (100) plane, and other two weak peaks, due to the (110) and (200) planes as shown in Fig. 3.2 . The slight shift of all three peaks to a smaller angle after impregnation SBA-15 cal. silica with zinc oxide, results of increasing the interplanar spacing(d)values Table3. 1. Obviously, the hexagonal ordered structure of silica are maintained well even after the mixing with metal precursor and calcination process. This provides evidence that the addition of inorganic precursor does not change the mesoscopic order of SBA-15 silica. This suggests that the insertion of zinc oxide is probably lead to expand the mesoporous silicate [93]. Only minor shift has been seen as the loading of copper oxide is increased from 5% to 20%.

Based on $a_0 = \frac{2\,d_{100}}{\sqrt{3}}$, where a_0 represents the pore-to-pore distance of the hexagonal structure, the unit cell parameter a_0 of the ZnO/SBA-15and CuO/SBA-15 nanocomposites are calculated to be increased from 10.1 for SBA-15 cal. Silica to 10.6 nm for metal oxide impregnated samples [92]. This of cause confirmed the expanding of silica network after loading of metal oxide precursors .

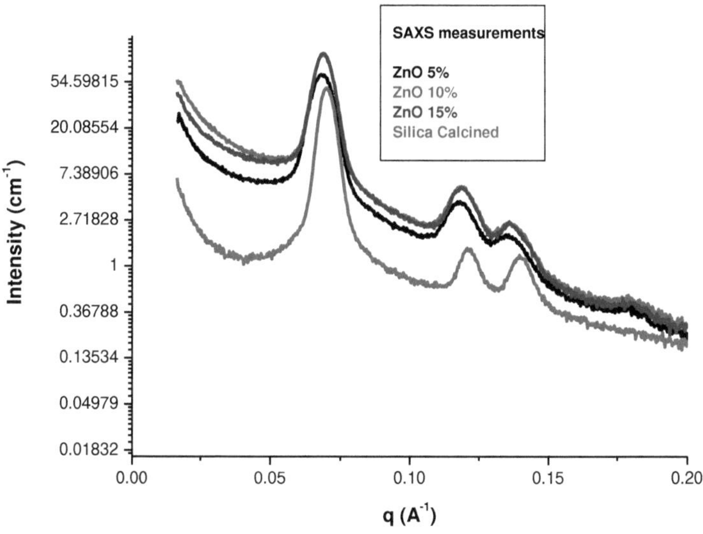

Fig. 3.3 SAXS of SBA-15 cal. and ZnO/SBA-15 composites at different percentages.

3.2.2 Wide Angle X-ray Scattering

Wide angle X-ray scattering patterns , of ZnO loaded into calcinated SBA-15 silica materials are displayed in Fig.3.4. The XRD patterns of embedded ZnO reveals that zinc oxide cluster is present in crystalline form probably into the pores of the silica network.   All the diffraction peaks are well indexed to the hexagonal ZnO wurtzite structure (JCPDS no. 36–1451) [94]. The major peaks at 2θ = 31.7, 34.4, 36.3, 47.6, 56.8, 62.9 can be indexed as (100), (002), (101), (102), (110), and (103) diffractions of the ZnO wurtzite [95], respectively. Therefore, ZnO in the ZnO/SBA-15 nanocomposite calcinated at 600 °C exists in form of the crystalline ZnO wurtzite phase. Reported results, showed that ZnO in the 20 wt% ZnO/SBA-15 nanocomposite calcinated at 500°C form a non-crystalline ZnO phase [96]. Diffraction peaks corresponding to the impurities were not found in the XRD patterns, confirming high purity of the synthesized composites. The mean crystallite size of ZnO particles was determined by Sherrer's equation (where $D = 0.89\lambda/\beta \cos\theta$ where D is the crystallite size (nm), λ is the wavelength of incident X-ray (nm), β is the full width at half maximum, and θ is the diffraction angle. The obtained particle size was in the range of 2.2 nm for the coated ZnO (Table3. 1).

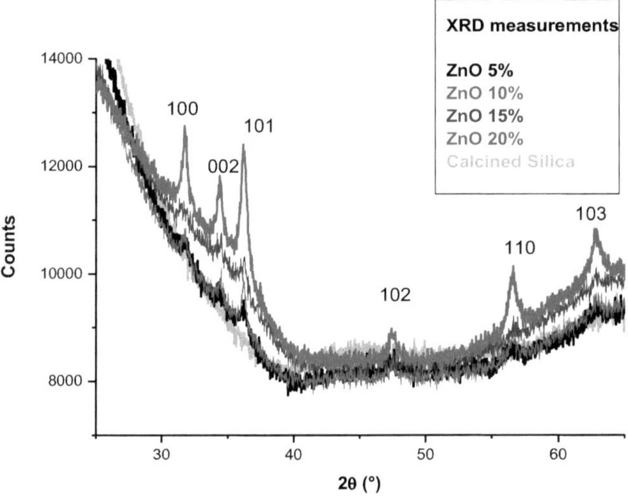

Fig. 3.4 X-RD analysis for ZnO/SBA-15 samples of different impregnation percentages of ZnO

XRD patterns of CuO/ SBA-15 cal. at different percentages figs3.5 reveals that copper oxide is present in crystalline form within the pores of the silica network. The patterns are correspond to the monolinic phase of CuO, the diffraction peaks match very well with the PDF file 80-1916 [93] . The peaks centered at $2\theta = 33.5$(sh), 35, 38.5, 48, 53 and 57° are assigned to the 110, -111, 111, -202, 020 and 202 reflection lines of monoclinic CuO particles. The peaks intensities are increased as loaded of copper oxide precursor increased from 5% to 20%. No peaks were observed for low content (5%) of copper oxide, probably the peaks have very low intensity and below the detection level. Scherer's equation was used to estimate mean crystallite size of nanoparticles. The average mean size of copper oxide nanoparticles estimated by XRD data was 2.1 nm (Table 3.1).

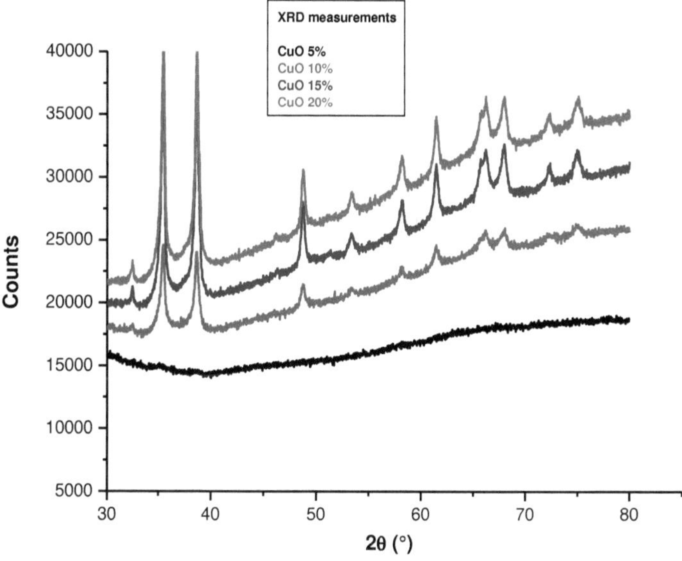

Fig. 3.5 XRD analysis for CuO/SBA-15 samples of different impregnation percentages of CuO.

3.3 FT - IR Spectra

FT-IR spectra of for pure SBA-15cal. and its impregnated ZnO/ SBA-15 (20%) composite are depicted in Fig. 3.6a & b, respectively. The two spectra are very similar , this means that the introduction of ZnO has no chemical interaction with the host SBA-15 silica. There is no peak correspond to the acetate around 1735 cm^{-1} in the IR spectra of ZnO/SBA-15 Figs.3.6b indicating complete decompose of zinc acetate by calcinations process and formation of ZnO cluster of particles. No new absorbance is observed in the modified silica composites, excluding the possibility that zinc oxide is inserted into the framework of SBA-15. The most significant difference is in the intensity of the two peaks at 3488 and at 1600 cm^{-1} of OH vibrations, which are due to

42

absorbed water molecules onto the surface of ZnO particles. This was also evident from TGA results. Abroad absorption band around 3488 cm^{-1} is correspond to the ν(O–H) group stretching vibrations of hydrogen bonded between H_2O molecules and ZnO crystal surface . The absorption band in the range of 1456–1642 cm-1 is probably due to δ(O–H) vibration. The formed ZnO phases is characterized by intense IR band which could be assigned to the Zn-O vibrations with poor resolved shoulders at about 500 cm^{-1}. The Si–O–Si stretching vibration bands at 1090 cm^{-1} and 812 cm^{-1} are due to asymmetrical and symmetrical stretching vibrations of Si-O bond ,respectively. The reduction in peak intensity at 500 cm^{-1} of Si–O peak after treatment with zinc oxide, may indicate presence of ZnO particles which were successfully inserted into the pores of silica network. One of the main features for the impregnation of ZnO particles is the high intensity of OH bands around 3500 due to absorbed crystallized water molecules in the ZnO crystals which was decreased at higher temperature as seen from TGA analysis.

Similar FTIR spectral pattern is obtainedfor pure silica SBA-15 and CuO/SBA-15 as shown in Fig. 3.7 a&b. These assignments are based to reported spectral data of similar systems [97-100].

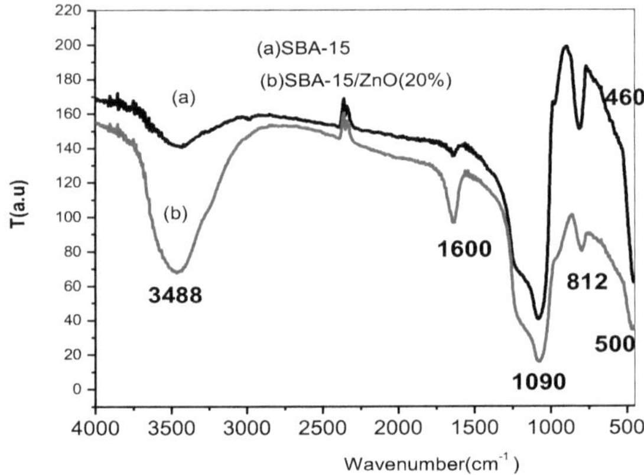

Fig. 3.6 FT-IR spectra for a) SBA-15cal.silica, b) ZnO (20%)/SBA-15

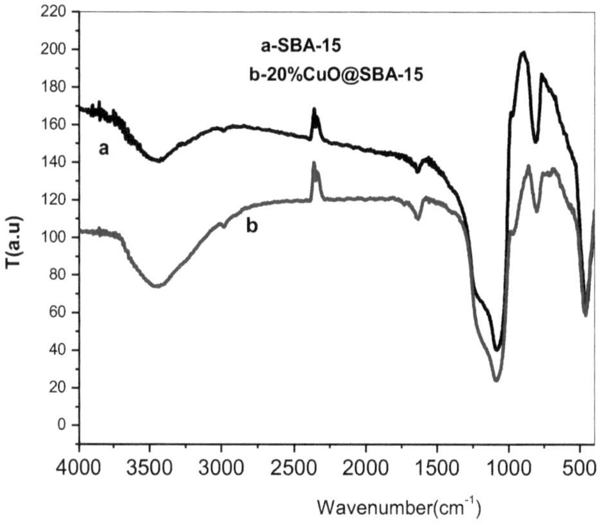

Fig. 3.7 FT-IR spectra for a) SBA-15cal. Silica, b)CuO(20%)/SBA-15

The IR spectra of ZnO/ SBA-15 and ZnO/ SBA-15-NH$_2$ are presented in Fig.3.8 a &b. The bands at 1089 and 812cm^{-1}are belong to the symmetrical and asymmetrical vibrations of the bond Si–O–Si, respectively. The band 463 and 967 cm^{-1}have been to the bending vibration of δ(Si–O)and stretching vibration ν (Si–OH) silanols, respectively. Thus the bands at 1089, 967, 812 and 463 cm^{-1}are assigned to SBA-15 framework [97,100]. Additionally, all samples show the band 1633 cm^{-1}which belongs to free molecularH$_2$O.Up on functionalization with amine silane there is a reduction of (O-H) vibration at 3500 cm^{-1} and 1640 cm^{-1}.

This assignment is consistent with the broad band centered at ca. 3440 cm^{-1} which is characteristic of physically absorbed molecular water interacting by H-bonding with surface silanol groups. Finally, in the spectra of ZnO/SBA15–NH$_2$, it is observed the characteristics bands that correspond to vibrations of the N–H bonds (1558, and 3270 cm^{-1}), and bands of the C–H bonds (2927 and 2882 cm^{-1}) of methylene groups. And a decreasing of intensity of the band at 960 cm^{-1} of Si-OH. These results confirm the successful functionalization of SBA-15 with monoamine silane coupling agent[97-100].

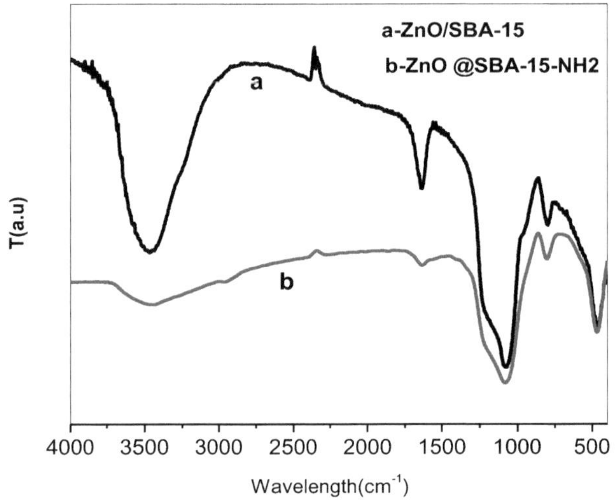

Fig. 3.8 FT-IR spectra for a) ZnO/SBA-15, b) ZnO/SBA-15-NH$_2$

3.4   TGA analysis

Thermo gravimetric analysis (TGA) and deferential thermo gravimetric analysis (DTA) for mesoporous SBA-15 silica, CuO/ SBA-15 silica, monoamine functionalized 20% CuO/ SBA-15 and monoamine functionalized free 20% CuO /SBA-15 silica were examined under nitrogen atmosphere at 20–600 °C at rate 10 °C/ minute. The thermogram of these materials are shown in Fig 3. 9 and Fig.3.10. Fig.3. 9-a shows two peaks, the main break occurs at ˜ 75°C due to loss of 1.3% of its initial weight. This attributed to loss of physisorbed water and alcohol from the system pores [101,102]. The second peak is at ˜390 °C due to loss of 1.4 %, which is probably due to dehydroxylation and loss of water or alcohol from silica [102,103]. The total loss of weight was 2.7 %.   Fig.3. 9-b shows the thermogram of the CuO/SBA-15 material, two peaks were observed, the first peak occurs at ˜ 75°C due to loss of 6.5 % of its initial weight. This is attributed to loss of a crystallized water molecules from trapped CuO into the system pores. The second peak at 350°C due to loss of 3.2% of its weight. This is attributed to dehydroxylation and loss of water from silica. The total loss of weight was 9.7 % . Figs.3. 9 (c &d) showed a total loss of weight was 14.2 due to loss of the amine functional groups in two steps as shown in Fig.3.10 at 300°C and 400°C. A dehydroxylation and loss of water or alcohol from the system was also observed at 450°C. These assignments were based on reported data of silica based systems[101,102] .

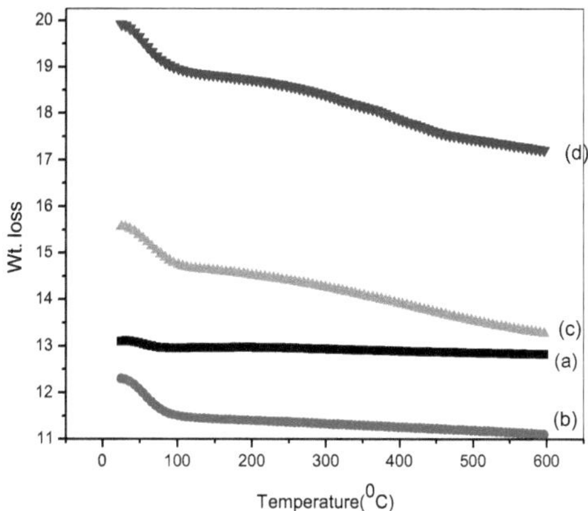

Fig. 3.9 TGA of a) SBA-15 silica , b) CuO/SBA-15, c) CuO/SBA-15-NH$_2$, d) free CuO/SBA-15-NH$_2$

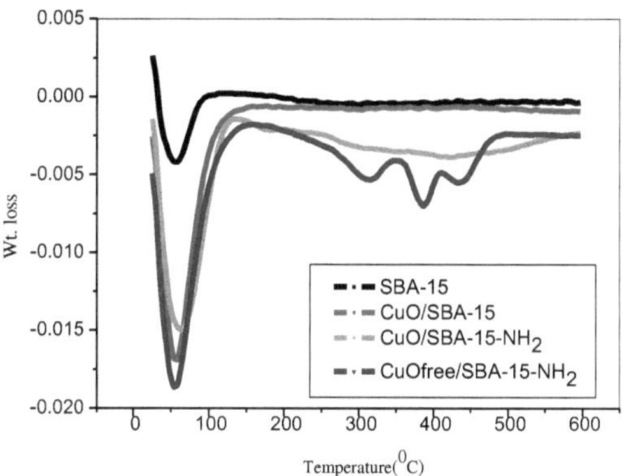

Fig. 3.10 DTA Curves of SBA-15 silica, CuO/ SBA-15, CuO/ SBA-15-NH$_2$ and free CuO/ SBA-15-NH$_2$

Thermogravimetric analysis (TGA) and deferential thermogravimetric analysis (DTA) for SBA-15 silica, ZnO/ SBA-15 silica, ZnO/SBA-15-NH$_2$ and free ZnO/ SBA-15 – NH $_2$ are given in Figs 3.11. (a-d) and Fig. 3.12 . Fig. 3.11a shows the thermogram of mesoporous SBA-15 silica, two peaks are observed, the first preak occurs at~75°C due to loss of 1.1% of its initial weight. This attributed to loss of physisorbed water and alcohol from the system pores. The second peak is at ~390°C due to loss of 1.2 %, which is probably due to dehydroxylation and loss of water or alcohol from silica. The total loss of weight was 2.3%. Fig.11b shows the thermogram of the ZnO/SiO$_2$ nanomaterial, two peaks were observed, the first preak occurs at ~75°C due to loss of 4.4 % of its initial weight. This is attributed to loss a crystalized water molecules from trapped ZnO of the system pores [84]. The second peak at 350°C due to loss of 4.4% of its weight. This is attributed to dehydroxylation and loss of water from silica. The total loss of weight was 8.8%. Figs 11. c& d showed three peaks at 75, 370 and 450 °C of a total loss of weight of 15 and 18 % respectively. These peaks are attributed to the loss of the amine organo functional groups, crystallized water molecules ,dehydroxylation and loss of water or alcohol from the system. These assignments are based on reported data of silica based systems[101,102].

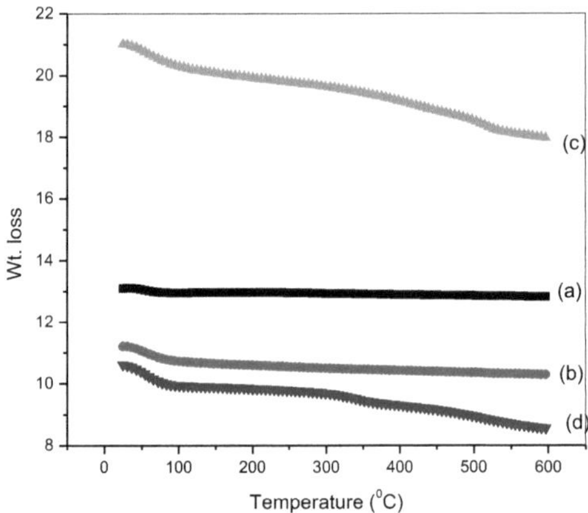

Fig. 3.11 TGA of a) SBA-15 silica, b) ZnO/SBA-15, c) ZnO/SBA-15-NH$_2$, d) free ZnO/SBA-15-NH$_2$

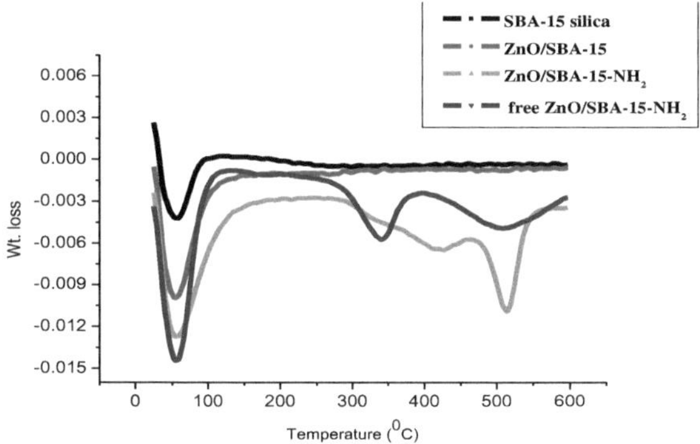

Fig. 3.12 DTA Curves of SBA-15 silica, ZnO/SBA-15 silica, ZnO/ SBA-15-NH₂and free ZnO SBA-15-NH₂

3.5 PL Spectra

It is commonly known that, ZnO exhibits the UV near-band-edge emission peak at around 380 nm, and the visible emission band ranging from 440 to 600 nm [104]. In general, the emission band in the visible region is associated with structural defects in ZnO [105]. The UV near-band-edge emission peak, is attributed to the recombination of free excitations [106], and depends on the ZnO particle size due to quantum size effect. The spectrum of ZnO/SBA-15 upon 325 nm excitation is given in Fig. 3.13. It can be seen from that, the PL spectrum ZnO/SBA-15 is dominated by emission band at 346 nm. Therefore, the emission band of ZnO/SBA-15 at 420 nm arises from the ZnO. This band is fitted by three peaks located at 420, 460, and 510 nm.

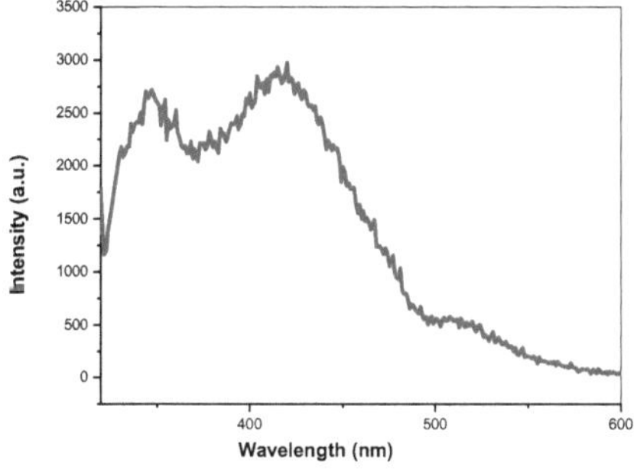

Fig. 3.13 PL spectra of ZnO/SBA-15

3.6 SEM and TEM analysis

SBA-15 silica material was analyzed by SEM and TEM techniques . Figs. 3.14 shows SEM image of high magnification power of SBA-15 material.It showed hexagonal particles of relative particle size of > 1.5 nm. The mesoporous ordered structure of SBA-15 material was confirmed from TEM analysis (Fig.3. 15 a & b).Its clear from TEM images of SBA-15 material that cylindrical channels appear along the 110 and 100 directions indicate of 2D p6mm mesostructure [107]. The estimated pore diameter is about 5.3 nm , center to center pore distance is about 10.6 nm which is exactly of same value obtained from SAXS analysis (Table 3. 1) , and pore wall thickness is about 5 nm are observed which are very closed to previously reported data [107] . One point BET analysis method of adsorption–desorption isotherms at 77K is used for determination of surface area for all materials. It is found that the calcinated SBA-15 has high surface area (900 m^2/g) than that of uncalcinated SBA-15 silica , 602 m^2/g (surfactant was washed with ethanol). This is probably that not all surfactant is removed.

Fig. 3.14 SEM image of SBA-15 cal.

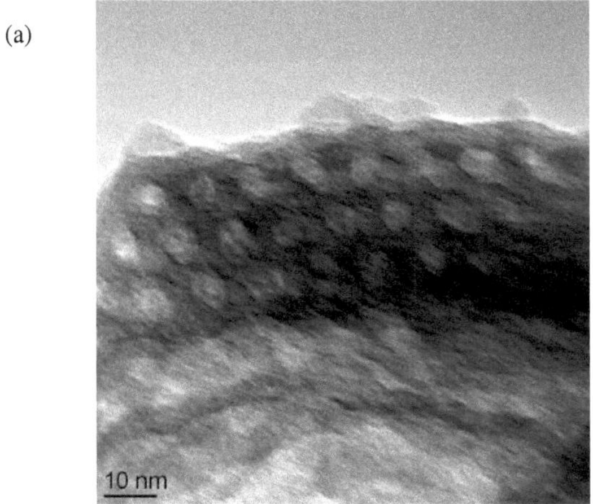

(a)

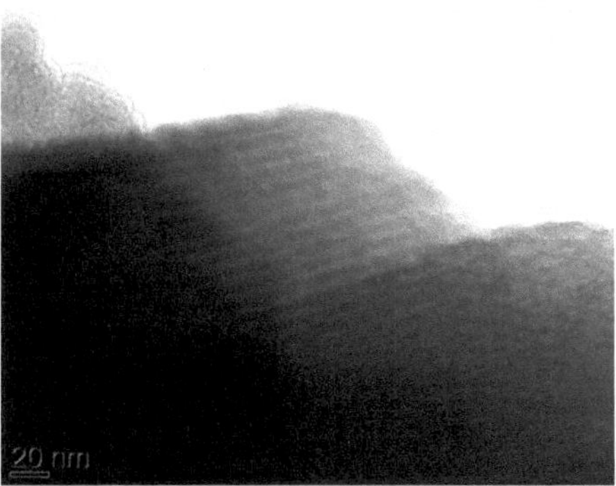

(b)

Fig. 3.15 TEM images of SBA-15 cal.

3.7 Application

3.7.1 E124 azo dye

E124 acid red is a synthetic azo dye. Tri sodium salt of 1-(4-sulpho-1-napthylazo)- 2-napthol- 6,8-disulphonic acid. It is molecular structure shown in Fig.3.16 ,used to induce a red color in a variety of food products like sweets, jellies, desserts, tinned and canned fruits and foods, cakes, pastries, soups and soft drinks.

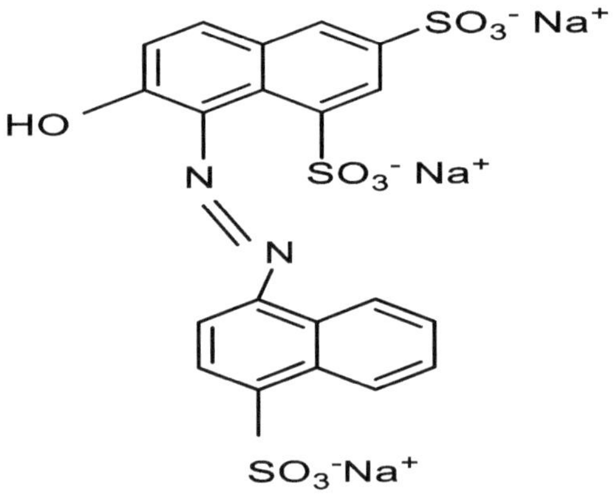

Fig. 3.16 E124 acid red azo dye chemical structure

There is no evidence of carcinogenicity, genotoxicity, neurotoxicity, or reproductive and developmental toxicity at the permitted dietary exposures; the European acceptable daily intake (ADI) is 0.7 mg/kg and the WHO/FAO ADI is 4 mg/kg.The possible ill effects or the possible side effects can be listed as follows:

- May elicit reactive responses in people who have an existing allergy to aspirin.

- It is also infamous for aggravating an existing asthma problem.

- May considered to be a carcinogenic substance in many countries.

- Responsible for hyperactivity in young children and kids [108].

3.7.2 Removal of E124 from aqueous solution

Treatment for drinking water production or water purification is the removal of contaminants from untreated water to produce drinking water that is pure enough for the most critical of its intended uses, usually for human consumption. Substances that are removed during the process of drinking water treatment include suspended solids, bacteria, fungi, minerals and toxic dyes.

In this work E124 dye was adsorbed by SBA-15-NH_2, ZnO/SBA-15-NH_2, free ZnO/SBA -15-NH_2.

When a functionalized amine materials is treated with E124 dye (70 ppm) solution, the white solid materials are readily changed into red, due to the entrapment of the E124 dye molecules into the pores and electrostatic attraction between the positive ammonium cations and (E124) dye anion (electrostatic attraction) as shown in Fig. 3 .17 .

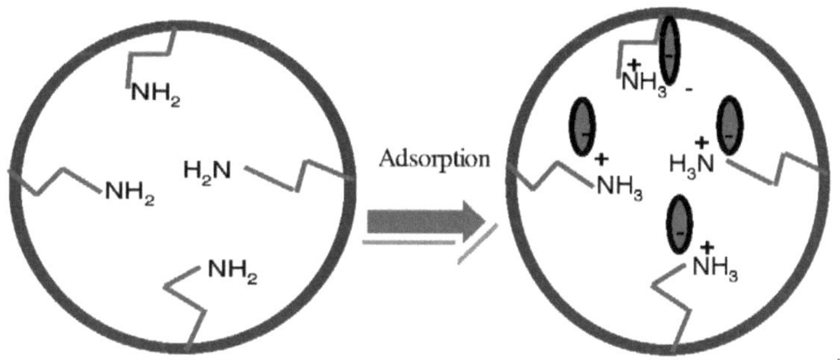

Fig.3.17 Adsorption of E124 dye

To make simple comparison between three functionalized amine adsorbents SBA-15-NH_2, ZnO/SBA-15-NH_2 ,free ZnO/SBA-15-NH_2 .The results show that ,free ZnO/SBA-15 functionalized amine is the best for removal E124 dye from water , the removal capacity was 69%.It is adsorbed the E124 dye quickly nearly after 10 minutes as shown in Fig. 3.16(a&b&c) .

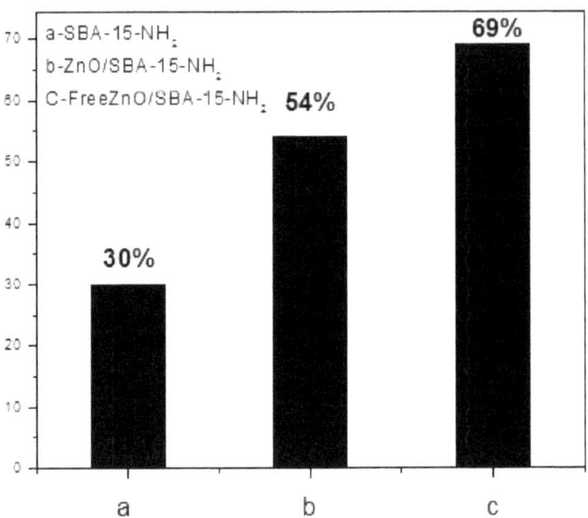

Fig. 3.17 adsorption capacity of a) SBA-15-NH$_2$, b) ZnO/SBA-15-NH$_2$, c) free ZnO/SBA-15-NH$_2$.

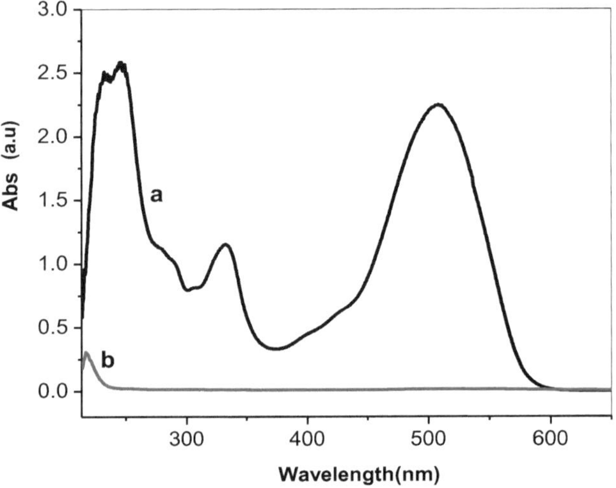

Fig. 3.18 UV-Vis absorption spectra of (a) E124 dye (b) free ZnO /SBA-15 – NH$_2$+E124

3.8 Conclusions

There were two methods for loading metal oxides into silica network, impregnation method and co-condensation method. In this work the copper oxide (CuO) and zinc oxide (ZnO) nanoparticles were inserted into SBA-15 silica network by impregnation method. SBA-15 silica was firstly prepared as free and fresh white powder by hydrothermal process using nonionic triblok co polymer surfactant (plourinc 123). It was soaked into the inorganic metal precursor solution of different percentages (5,10,15and 20 %) ,then the mixture was heated at 80 ^{0}C to dryness then heated at 100^{0}C ,followed by calcination at 600^{0}C for 5 h .

Several methods and techniques were used to examine their structure ,including FT-IR, XRD diffraction (SAXS and WAXS) ,TGA and PL spectra ,the following conclusions were considered:

1- CuO/SBA-15 and ZnO/SBA-15 silica with different percentages of metal oxide (5,10,15, and 20%) were prepared.

2- Amine functionalized CuO/SBA-15-NH$_2$ or ZnO/SBA-15-NH$_2$ composites were also prepared by treatment of metal oxide coated silica composites with monoamine silane coupling agent.

3- FT-IR spectra of the metal oxide coated SBA-15 silica and that of pure SBA-15 silica are very similar, which indicate that the inorganic precursor were not chemically bound to the silica network and probably physically bonded with silanols groups within the silica pores.

4- The mesoporouse structure of SBA-15 silica was retained even after impregnation with inorganic metal oxide of different percentages with slight shift to smaller diffraction angle.

5- XRD analysis proved that the metal oxide loaded in to the SBA-15 silica are in the crystalline form and the increasing of percentage of the inorganic oxide precursor don't change the X-RD pattern , but increasing its peak intensity .

6- TGA, FT-IR spectra proved that , the amine functionalized groups were successfully grafted into the surface impregnated metal oxide SBA-15 silica .

7- Adsorption of E124dye in the SBA-15-NH$_2$, ZnO/SBA-15-NH$_2$, free ZnO/SBA -15-NH$_2$ were examined . The results show that ,free ZnO/SBA-15 functionalized amine is the best for removal E124 dye from water , the removal capacity was 69%. It is adsorbed the E124 dye quickly.

Finally, we can say that, impregnation method is very good method to incorporate metal oxides into pores of mesoporous silica SBA-15.That because we can obtain the mesoporous silica in the solid state in the first step, so we find that, the mesoscopic hexagonal order of SBA-15 doesn't change after metal oxides loading.

In another hand, grafting propyl amine by post method, is very active process to improve adsorption capacity of mesoporous silica SBA-15 for water purification.

References

References

[1] C. kresge, M. leonowicz , W. Roth, J. Vartulli, J. Bck. J. Nature 1992, 359, 710.

[2] S.Tangestaninejad, M.Moghadam, V.Mirkhani, I.Baltorkand K. Ghani, J. Catal. Commun. 2009, 10, 853

[3] Z. AlOthman, A. Apblett, J. Mater. Lett. 2009, 6, 2331.

[4] K.Parida, S. Dash, J. Mol. J. Appl. Catal. A 2009, 306, 54.

[5] Z.AlOthman , A.Apblett. J. Surf. Sci. 2010, 256, 3573.

[6] M. Davis, R. Lobo. J. Chem. Mater. 1992, 4, 756.

[7] B.Lok, T.Cannon, C.Messina . J. Zeol. 1983, 3, 282.

[8] A. Sayari. J.Stud. Surf. Sci.and Catal. 1996, 102, 1.

[9] K. Song, J.Guan, Z. Wang, Q.Kan. J. Appl. Surf. Sci. 2009, 255, 5843.

[10] J.Beck, J.Vartuli, W.Roth, M.Leonowicz, K.Schmitt, D.Olson. J.Am. Chem.Soc. 1992, 114, 10834.

[11] C. Chen, S .Burkett, M. Davis. J. Microp. Mater. 1993, 2, 27.

[12] A.Monnier, F. Schüth, Q. Huo, D.Kumar, D.Margolese, R.Maxwell, G.Stucky, M. Krishnamurty, P. Petroff, A. Firoouzi, M.Janicke, B. Chmelka, J. Science 1993, 261,1299.

[13] A. Steel, S.Carr, M. Anderson. J. Chem. Soc. Chem. Commun. 1994, 13, 1571.

[14] J.Broekhoff. J. Stud. Surf. Sci. and Catal. 1979, 3, 663.

[15] X. Zhao, G. Lu, G.Millar. J. Chem. Res. 1996, 35, 2075.

[16] P.Tanev, T.Pinnavaia, J.Science 1995, 267, 865.

[17] M. Lawrence. J. Chem. Soc. Rev. 1994, 23, 417.

[18] P. Fromherz, J. Chem. Phys. Lett. 1981, 77, 460.

[19] D. Myers. Surfactant Science and Technology, 2end.Ede. VCH: New York, NY, USA, 1992,

[20] A.Cheetham, C. Mellot, J. Chem. Mater. 1997, 9, 2269.

[21] W. Stevens, K.Lebeau, M. Mertens, E.Vansant. j.Phys.Chem. 2006,110,9183.

[22] F.zhang ,Y.Yan, H. Yang, Y. Meng. J.Phys.Chem.2005,109 ,8723.

[23] Q.Huo, D.Margolese, U.Ciesla, P.Feng. J. Nature1994, 368, 317.

[24] D. Zhao, J. Sun, Q. Li, G. Stucky, J. Chem. Mater. 2000,12, 275

[25] T.Sun, J. Ying, J. Nature1997, 389, 704.

[26] C. Landry, S. Tolbert, K.Gallis, A.Monnier, G. Stucky, F.Norby, J. Hanson, J. Chem. Mater. 2001, 13, 1600.

[27] C.Brinker, G. Scherer, Sol-Gel Science. The Physics and Chemistry of Sol-Gel Processing; Edi. Academic Press: New York, 1990.

[28] W.Stöber, A.Fink, E.Bohn, J. Coll. and Interf. Sci. 1968 ,26, 62.

[29] N. Mal, M. Fujiwara, Y. Tanaka, J. Nature 2003, 421, 350.

[30] R.Feynman. J. Coll. and Interf. Sci. 1991, 254, 1300.

[31] L.D´Souza, R.Richards, J.A.Rodríguez, Fernández-García, Synthesis of Metal-Oxide Nanoparticles: Liquid-Solid transformations in "Synthesis, Properties and Applications of Oxide Nanoparticles" M; Edi. Whiley: N. J., 2007. Chpt. 3.

[32] K. Suslick, S. Choe, A. Cichowlas, M. Geenstaff, J. Nature 1991, 353,414.

[33] L. Interrante, M. Hampen-Smith, Chemistry of Advanced Materials, Edi. VCH, Wiley, VCH, 1998.

[34] V.Uskokovick, M.Drofenik, J. Surf. Rev. Lett.2005, 12, 239.

[35] M. Ohring, Material Science of Thin Films, Edi. Academic-Press: San Diego,1992

[36] G.Hubler, J. Mater. Res. Bull. 1992, 17, 25.

[37] B. Sun , H. Sirringhaus, J. Nano. Lett. 2005, 5, 2408.

[38] R. Konenkamp, R. Word ,M. Godinez, Nano. Lett., 2005, 5, 2005.

[39] J. Wang, X. Sun, Y. Yang, H. Huang, Y. Lee, O. Tan, L.Vayssieres, J. Nanotech. 2006, 17, 4995.

[40] M. Height, S. Pratsinis, O. kasuwandumrong, P. Praserthdam, J. Appl. Catal., 2006, 63, 305.

[41] A. Becheri, M. Durr, P. Lo Nostro, P. Baglioni, J. Nanopart. Res., 2008, 10, 679.

[42] A. Rakhshni, J. Sol. Sta. Elec. 1986, 29, 7.

[43] K. Klabunde, J.Stark, O.Koper, C.Mohs, D.Park, S.Decker, Y. Jiang, I. Lagadic, D.Zhang. J.Phys.Chem.1996 ,100,12142.

[44] R. Richards , W. Li , S. Decker , C. Davidson , O. Koper , V. Zaikovski , A. Volodin , T. Rieker , K. Klabunde .J. Chem. Soc. 2000,122,4921.

[45] A. Balanta, C. Godard, C. Claver, J. Soc. Rev. 2011, 40, 4973.

[46] X. Zhao, G. Lu, A. Whittaker, G. Miller, H. Zhu, J.Phys. Chem. 1997 101,6525.

[47] L. Mercier, T. Pinnavaia, J. Chem. Mater. 2000, 12,188.

[48] D. Macquarrie, D. Jackson, J. Mdoe, J. Clark, J. New J.Chem., 1999 ,23,539.

[49] A. Stein, B. Melde, R. Schroden, J. Adv. Mater. 2000,12, 1403.

[50] M. Espinosa, S. Pacheco, S. Vargas, M. Estevez, M.E. Llanos, R. Rodríguez, J.Appl. Catal. A 2011, 401,119.

[51] C. Tu, A. Wang, M. Zheng, X. Wang, T. Zhang, J. Appl. Catal. A 2006, 297, 40.

[52] J. Deng, L. Zhang, H. Dai, Y. Xia, H. Jiang, H. Zhang, H. He, J. Phys. Chem. 2010, 114, 2694.

[53] K. Suslick, S. Choe, A. Cichowlas, M. Grinstaff, J. Nature 1991, 353,414.

[54] A. Gedanken, X. Tang, Y. Wang, N. Perkas, Y. Koltypin, M. Landau, L. Vradman,M. Herskowitz, J.Chem. Eur. 2001, 7,4547.

[55] H. Li, R. Wang, Q. Hong, L. Chen, Z. Zhong, Y. Koltypin, J. Calderon, A.Gedanken, J.Ame. Chem. Soci. 2004, 20,8352

[56] S. Tangestaninejad, V. Mirkhani, M. Moghadam, I. Mohammadpoor, E.Shams, H. Salavati, J. Ultrason. Sonochem. 2008, 15, 438.

[57] X. Zhang, N. Huang, G. Wang, W. Dong , M. Yang, Y. Luan, Z.Shi, J. Microp. and Mesop. Mater. 2013,177, 47.

[58] Y. Wei, Y.Cao, J. Zhu. Stud. in Surf. Sci. and Catal. 2004,154, 878.

[59] Y. Sun, S. Walspurger, J. Tessonnier, B. Louis, J. Sommer. Appl. Catal. A. 2006, 300, 1.

[60] P. Shah, A. Ramaswamy, R.Pasricha, K. Lazar, Ramaswamy, J.Stud. in Surf. Sci. and Catal. 2004, 154 ,870.

[61] L.Baitao, S. Zhang J. Intern. Jour. Hydr. Ener. 2013, 38,14260.

[62] J. Raoofa, F. Chekinb, V. Ehsania, J. Sens. and Actu. 2015,207, 291.

[63] D. Guoan, L. Sangyun, P. Mathieu, W. Chuan, F. Fang, P. Lisa, L. Gary, J. of Catal. 2008, 253,74.

[64] L. Chmielarz, P. Kus´trowski, R. Dziembaj, P. Cool, E. Vansant, J. Microp. and Mesop. Mate. 2010, 127, 133.

[65] N. Wanga, X. Yu, K. Shen, W. Chu, W. Qiana, J. Inter. Jour. Hydr. Ener. 2013, 38, 9718.

[66] G. D. Mihai, V. Meynen, M. Mertens,N. Bilba, P. Cool, E. F.Vansant, J.Mater.Sci. 2010,45,5786.

[67] A. Chen ,W. Zhang ,X. Li , D. Tan ,X. Han. J.Catal .Lett .2007, 119, 159.

[68] D. Zhao, J. Feng, Q. Huo, B. Chmelka, G. Stucky, J. Amer. Chem. Soc., 1998, 279, 548.

[69] W. Zhengying, W. Yimeng, Z.Jianhua . J.Chin. Sci. Bull. 2004, 49, 1332.

[70] K. Moller, T. Bein, J. Chem. Mater. 1998,10, 2950.

[71] A. Stein, B. Melde, R. Schroden, J.Adv. Mater. 2000 ,12,1403

[72] A. Sayari, S. Hamoudi, J. Chem. Mater. 2001,13, 3151.

[73] M.Puanngam, F.Vnob. J.Haza.Mate.2008,154,578.

[74] Z.Dan, L.Hua. J.Chin.Sci.Bull.,2013,58,879.

[75] A. Chong, X. Zhao, A. Kustedjo, S. Qiao. J. Micr. and Meso. Mate. 2004,72,33.

[76] N. Velikova, Y. Vueva, Y. Dimitriev, I. Miranda, J. Inter. Jour. Mater. Chem. 2013,3,21.

[77] J. Wang, T. Tsuzuki, L. Sun, X. Wang . J. Am. Ceram. Soc. 2009,92 ,2083.

[78] S. Hsu. , Ph.D. Infrared Spectroscopy. Separation Sciences Research and Product Development Mallinckrodt,1992,Chp.15,p249

[79] Skoog. Principles of Instrumental Analysis2007. 6th ed. p. 169.

[80] J. Garcia, J. Hernandez, J. Alvarez, E. Cruz, G. Puente, J. Appl. Phys. 1990, 67, 3810.

[81] A.Coats, J.Redfern, J. Anal. (1963),88,906 .

[82] J. Brady, B. Shelby. J. Geol. Edu. 1995 ,43, 471.

[83] www.biosaxs.com /technique.html

[84] Q. Jiang, Z. Wu, Y. Wang, Y.Cao, C. Zhou, J. Zhu, J. Mater. Chem. 2006, 16, 1536.

[85] G. Tang, Y.Xiong, L. Zhang, G. Zhang, J. Chem.Phys.Lett. 2004,395, 97.

[86] J.Chen, Z. Feng, P. Ying, C. Li, J. Phys. Chem. 2004, 108,12669.

[87] W.Zeng, Z.Wang, X. Qian, J.Yin, Z. Zhu, J. Mater. Res. Bull. 2006, 41, 1155.

[88] W. Zhang, J. Shi, L. Wang, D. Yan, J. Chem. Mater. 2000,12,1408.

[89] H.Chen,J.Shi,H.Chen,J.Yan,Y.Li,Z.Hua,J .Yang,D.SYan. J.Opt.Mater.2004,25, 79.

[90] L. Qingshan, Z. Wang, L.Jiangong, P. Wang, X. Ye, J. Nanos. Res. Lett. 2009, 4,646.

[91] D.Zhao, Q.Huo, J. Feng, B. Chmelka, G. Stucky, J. Am. Chem. Soc. 1998,120, 6024.

[92] D.Zhao, J.Feng, Q.Huo, N.Melosh, G. Frederickson, B. Chmelka, G. Stucky. J. Science. 1998,279,548.

[93] Y. Wang, Z. Wu, L. Shi, J. Zhu, J. Adv. Mater 2005, 17, 323.

[94] Q. Lu, G. Yun, Q. Tran, W. Zaho, J. Mater. Lett. 2013,93, 12.

[95] K.F.Lin, H.M.Cheng,H.C. Hsu,L.J. Lin,W.F. Hsieh. J. Chem.Phys. Lett.2005,409, 208.

[96] S.Vepřek, Z.Iqbal, F.A.Sarott, J. Philos. Mag. 1982,45,137.

[97] M. Ramasamy, Y. Kim, H. Gao , D. Yi , J. An, J. Mater. Rese. Bull. 2014, 51,85.

[98] J.Sauer,F.Marlow, B.Spliethoff, J. Chem.Mater.2002,14,217.

[99]   C. Halliwell, A. Cass, J.Anal. Chem. 2001, 73, 2476.

[100] P. Lihitkar , S. Violet, M. Shirolkar, J. Singh, O. Srivastava, R. Naik, S. Kulkarni, J. Mate. Chem. and Phys. 2012, 133, 850.

[101] W. Zhengying, W. Yilun, W. Yimeng, Z. Jianhua, Chin. Scie. Bull. (2004) 49,1332.

[102] Y. S. Li , J. S. Church, A. L. Woodhead, F. Moussa,SpectrochimicaActa Part 2010,76 , 484.

[103] I. El-Nahhal, J. Salem, S. Kuhn, T. Hamad, R. Hempelmann, S. Al Bhaisi, J.Powd. Tech. 2015.287,439.

[104] C. Huang , K. Chang , H. Oua, Y. Chiang, C. Wang, 2011,141, 102.

[105] S.Cho, J. Ma, Y. Kim,Y. Sun, G. Wong, J. Ketterson, J. Appl. Phys. Lett. 1999,75, 2761.

[106] Ü.Özgür, Y.Alivov, C. Liu, A.Teke, M.Reshchikov, S.Doğan,V.Avrutin, S. Cho, H. Morkoç, J. Appl. Phys. 2005, 98, 041301.

[107] E. Wong, P.C. Searson. J. Appl. Phys.Lett. 1999, 94, 2939.

[108] F.Aguilar , U.Rcharrondier, B.Dusemund. J.Euro.Foo.Saf.Auth.2009, 7,39.

Printed by Books on Demand GmbH, Norderstedt / Germany